Six Degrees

THE SCIENCE OF A
CONNECTED AGE

Duncan J. Watts

W. W. NORTON & COMPANY

NEW YORK LONDON

Copyright © 2003 by Duncan J. Watts
Saul Steinberg's "View of the World from 9th Avenue," © 2002 The Saul Steinberg
Foundation/Artists Rights Society (ARS), New York.

For information about permission to reproduce selections from this book, write to
Permissions, W. W. Norton & Company, Inc., 500 Fifth Avenue, New York, NY 10110

Manufacturing by The Haddon Craftsmen, Inc.
Book Design by Barbara M. Bachman
Production manager: Andrew Marasia

LIBRARY OF CONGRESS CATALOGING-IN-PUBLICATION DATA

Watts, Duncan J., 1971–
Six degrees : the science of a connected age / Duncan J. Watts.— 1st ed.
p. cm.
Includes bibliographical references and index.
ISBN 0-393-04142-5 (hardcover)
1. Graph theory. 2. Network analysis (Planning) 3. Social networks—
Mathematical models. I. Title.
QA166 .W37 2003
511'.5—dc21

2002013331

W. W. Norton & Company, Inc., 500 Fifth Avenue, New York, N.Y. 10110
www.wwnorton.com

W. W. Norton & Company Ltd., Castle House, 75/76 Wells Street, London W1T 3QT

1 2 3 4 5 6 7 8 9 0

SIX DEGREES

For

Mum and Dad

"I read somewhere that everybody on this planet is separated by only six other people. Six degrees of separation. Between us and everybody else on this planet."

—*Ouisa, in*
SIX DEGREES OF SEPARATION,
by John Guare

Contents

Preface

*"I rarely end up where I was intending to go, but often
I end up somewhere that I needed to be."*
—DOUGLAS ADAMS,
The Long Dark Tea-Time of the Soul

IT'S FUNNY HOW THINGS WORK OUT. IT HAS BEEN LESS THAN A
decade since I stared down that long corridor at Cornell, wondering
why I'd moved halfway around the planet to come study some obscure
subject in a place that all of a sudden looked like a prison. Yet in that
brief time, the world has changed several times over, and my world
with it. Surprised by the meteoric rise of the Internet, stung by a series
of financial crises from Asia to Latin America, and stunned by ethnic
violence and terrorism from Africa to New York, the world has learned
the hard way that it is connected in a manner few people had anticipated and no one understood.

In the quiet corridors of academia, meanwhile, a new science has
been emerging—one that speaks directly to the momentous events
going on around it. For want of a better term, we call this new science
the *science of networks*. And unlike the physics of subatomic particles or
the large-scale structure of the universe, the science of networks is the
science of the real world—the world of people, friendships, rumors,
disease, fads, firms, and financial crises. If this particular period in the
world's history had to be characterized in any simple way, it might be as

one that is more highly, more globally, and more unexpectedly connected than at any time before it. And if this age, *the connected age,* is to
be understood, we must first understand how to describe it scientifi-
cally; that is, we need a science of networks.

This is a story about that science. It isn't *the* story, the unabbrevi-
ated version being already far too large to accommodate in one small
volume and soon beyond the ability of any one person to learn in a lifetime. Rather, it is a fragment, a single traveler's tale of his journeys in a
strange and beautiful land. In any case, *every* story must be told from
some perspective (whether it is done so openly or not), and this one is
told from mine. In part that's because I played a role in the events
themselves, events that have been central to the trajectory of my own
career. But there is another, deeper reason, one that has to do with the
telling of science. The science of textbooks is typically a dry and intimidating affair. Unfolding in a relentless march of logic from apparently
impossible questions to seemingly indisputable conclusions, textbook
science is hard enough to follow, let alone emulate. And even when science is presented as an act of discovery, an achievement of humans, the
process by which they actually figured it out remains cloaked in mystery. My dominant memory from years of physics and math courses is
the depressing sense that no normal person could actually do this stuff.

But real science doesn't work that way. As I eventually learned, real
science occurs in the same messy ambiguous world that scientists
struggle to clarify, and is done by real people who suffer the same kind
of limitations and confusions as anybody else. The characters in this
story are, one and all, talented people who have worked hard throughout their lives to succeed as scientists. But they are also entirely human.
I know that because I know them, and I know that we have struggled
and often failed together, only to pick ourselves up to try again. Our
papers get rejected, our ideas don't work out, we misunderstand things
that later seem obvious, and most of the time we feel frustrated or just
plain stupid. But we struggle on, the journey being every bit as much
the point as the destination. Doing science is really a lot like doing any-

thing else, but by the time it gets out into the larger world and everyone reads about it in books, it has been so reworked and refined that it takes on an aura of inevitability it never had in the making. This story is about science in the making.

Of course, no story takes place in a vacuum, and one thing that I hope to convey in this book is a sense of where the science of networks comes from, how it fits into the larger scheme of scientific progress, and what it can tell us about the world itself. There is far more to say about these matters actually than I can include here, because rather a lot of people have been thinking about networks for rather a long time. But as much as the story omits (and it omits an awful lot), I hope it will convey the point that the connected age cannot be understood by trying to force it into any one model of the world, however reassuring that might seem, nor can it be understood by any one discipline working alone. The questions are simply too rich, too complicated, and, frankly, too hard for that.

Equally frankly, the science of networks doesn't have the answers yet either. As tempting as it is to overstate the significance of our findings, the truth is that most of the actual science here comprises extremely simple representations of extremely complicated phenomena. Starting off simple is an essential stage of understanding anything complex, and the results derived from simple models are often not only powerful but also deeply fascinating. By stripping away the confounding details of a complicated world, by searching for the core of a problem, we can often learn things about connected systems that we would never guess from studying them directly. The cost is that the methods we use are often abstract, and the results are hard to apply directly to real applications. It is a necessary cost, unavoidable in fact, if we truly desire to make progress. Before engineers could build airplanes, physicists first had to understand the fundamental principles of flight, and so it is with networked systems. In the pages ahead, we will speculate about the promising applications of the simple network models—we will try to imagine what the magnificent flying machines might eventually look like. But at the end of the day we must be honest and distinguish speculations from the state of the science itself. So if you're looking for answers, try the new-age section. Only by delineating what it can explain from what it cannot does science become powerful, and theories that confuse the two ultimately do us a disservice.

What the science of networks *can* do, even now, is give us a different way to think about the world, and in so doing help us to shed new light on old problems. To that end, this book is actually two stories in one. First, it's a story about the science of networks itself—where it came from, what has been figured out, and how it was figured out. And second, it's a story about phenomena in the real world, such as epidemics of disease, cultural fads, financial crises, and organizational innovation, that the science of networks attempts to understand. Both these stories run in parallel throughout the book, but some chapters emphasize one over the other. Chapters 2 through 5 are mostly about different approaches to understanding real-world networks, how the various academic disciplines have contributed to the discovery process, how my own involvement began through my work with Steve Strogatz on small-world networks, and how that work has developed and expanded in the years since. Chapters 6 through 9 focus more on a networked *way of thinking* about the world, and its application to problems like disease spreading, cultural fads, and business innovation, rather than treating the networks themselves as the objects of study.

Although each chapter builds on its predecessors, it is not necessary to read everything from start to finish. Chapter 1 sets up the context of the story, and chapter 2 fleshes out its background. If you wish to skip these sections and proceed straight to the new science, you may do so (you'll be missing out though). Chapters 3, 4, and 5 more or less go together, describing the creation and implications of various models of networked systems, especially the so-called small-world and scale-free network models on which much recent work has been done. Chapter 6 discusses the spreading of diseases and computer viruses and can be read with limited references to the previous chapters. Chapters 7 and 8 deal with the related but distinct subject of social contagion, and what it tells us about cultural fads, political upheavals, and financial bubbles. Chapter 9 discusses organizational robustness and its lessons for modern business firms. And chapter 10 wraps up the story, providing a brief overview of the state of play.

Like the stories it relates, this book has a history, one that involves quite a number of people. Over the past few years, my collaborators and colleagues—especially Duncan Callaway, Peter Dodds, Doyne Farmer, John Geanakoplos, Alan Kirman, Jon Kleinberg, Andrew Lo, Mark Newman, Chuck Sabel, and Gil Strang—have been a constant source of ideas, encouragement, energy, and entertainment. It would have been hard to write this book without them because there wouldn't have been much to write about. Even the best subject matter, however, is not enough. Without the encouragement of Jack Repcheck at Norton, and Amanda Cook at Perseus, I would never have gotten started. And without the gentle guidance of Angela von der Lippe, my editor at Norton, I never would have finished it. My thanks also go to the generous souls— Karen Barkey, Peter Bearman, Chris Calhoun, Brenda Coughlin, Priscilla Ferguson, Herb Gans, David Gibson, Mimi Munson, Mark Newman, Pavia Rosati, Chuck Sabel, David Stark, Chuck Tilly, Doug White, and especially Tom McCarthy—who volunteered to read and comment on various drafts. Gueorgi Kossinets provided invaluable assistance in preparing the many figures, and Mary Babcock did a fantastically thorough job of copy editing.

At a more general level, I am deeply grateful to a number of people at Columbia University—Peter Bearman, Mike Crowe, Chris Scholz, and David Stark—as well as Murray Gell-Mann, Ellen Goldberg, and Erica Jen at the Santa Fe Institute and Andrew Lo at MIT for giving me the freedom and support to pursue my selfish interests, even sometimes at questionable benefit to their own. The National Science Foundation (under grant 0094162), Intel Corporation, the Santa Fe Institute, and the Columbia Earth Institute have provided critical financial support to my teaching and research, as well as to a series of seminal workshops in Santa Fe and New York, out of which numerous collaborations and projects have sprung. But among the multitude of influences, both institutional and personal, from which I have benefited, there are two who stand out. The first is Steve Strogatz, who over the years has been an inspirational mentor, an invaluable collaborator, and a good friend. And the other is Harrison White, who first brought me to Columbia, first connected me with the Santa Fe Institute, and ultimately brought me into sociology. Without these two, as they say, none of this would have been possible.

And finally, my parents. It is probably foolish to speculate about the influence a person's upbringing has on the course of his or her life, but in my case a few things seem clear. My father, the first scientist I ever knew, was also the first person to guide me through the pleasures and pains of original research, and in his own way stimulated the entire thought process out of which this book grew. My mother, meanwhile, not only taught me how to write but also impressed on me at an early age that ideas only realize their power when people understand them. And together, through the example of their own quietly remarkable lives, they gave me the courage to try things that I never really believed could work out. This book is for them.

DUNCAN WATTS
NEW YORK CITY
MAY 2002

CHAPTER ONE

The Connected Age

THE SUMMER OF 1996 WAS A SIZZLER. ACROSS THE NATION, THE
mercury was climbing to record highs and staying there, a mute testi-
mony to climatic unpredictability. Meanwhile, closeted in their
domestic fortresses, Americans were stacking their refrigerators,
cranking the air-conditioning, and no doubt watching a record amount
of mind-deadening television. In fact, no matter what the season or the
weather, Americans have become increasingly reliant on a truly staggering and ever growing array of devices, facilities, and services that
have turned a once hostile environment into the lifestyle equivalent of a
cool breeze. No amount of inventiveness or energy is excessive if it
results in the creation of leisure, the increase of personal freedom, or
the provision of physical comfort. From climate-controlled vehicles the
size of living rooms to climate-controlled shopping malls the size of
small cities, no effort or expense has been spared in modern America's
endless crusade to impose strict discipline on a once unruly and still
occasionally uppity planet.

Driving this relentless engine of civilization is an entity as mun-
dane and familiar as the landscape itself, yet as profoundly life altering
as any of mankind's inventions—the power system. Stretched like a
spider's web across the entire North American continent lies an enor-

mous network of power stations and substations and the high-voltage transmission cables that connect them. Drooping near trees along country lanes, straddling the steep ridges of the Appalachians, and marching like columns of giant soldiers across the endless plains of the West, the electric power transmission grid is at once the life blood of the economy and the soft underbelly of civilized life.

Built at vast expense over the better part of the last century, the power system is arguably the most essential technological feature of the modern world. More pervasive even than highways and railroads, and more fundamental than cars, airplanes, and computers, electric power is the substrate onto which every other technology is grafted, the foundation for the grand edifice of the industrial and information ages. Without power, pretty much everything we do, everything we use, and everything we consume would be nonexistent, inaccessible, or vastly more expensive and inconvenient. Electricity is a fact of life so basic that we cannot imagine being without it, and if we are forced to do so, it can be tremendously destructive in the most primal fashion, as New York City discovered over twenty-five awful hours in 1977. Back then, in a society that had barely discovered computers and whose automobiles, factories, and household contraptions were far less dependent on electronics than they are today, a blackout resulting from a virtually unforeseeable combination of small mistakes and systemic weaknesses plunged New York into darkness, and its nine million inhabitants into a mayhem of riots, plundering, and widespread panic. When the lights came back on and the debris had been cleared, the bill for damages ran to some $350 million. The catastrophe so alarmed politicians and regulators alike that they vowed not to let it happen again and implemented a number of stringent measures to seal their promise. As we have since discovered, in a complex connected world, even the best-laid plans can be little more than the proverbial rearrangement of deck chairs on the *Titanic*.

Like every infrastructure, from highway systems to the Internet, the power grid is not really a single entity, but several regional networks cobbled together under the rubric of greater connectivity for the good of all. The largest of these administrative units is the collection of roughly five thousand stations and fifteen thousand lines that make up the transmission grid of the Western Systems Coordinating Council, a conglomeration of power generators and distributors that is responsible for providing electricity to everyone and everything west of the Rocky Mountains, from the Mexican border to the Arctic. In the blistering heat of August 1996, everyone with an air-conditioner was running it full blast and every ice-cold Budweiser at every backyard barbecue was drawing its share of power from the grid. The summer tourist crowds, reluctant to head back east, were lingering in the coastal cities, temporarily swelling the already bulging populations of Los Angeles, San Francisco, and Seattle and straining the aging and already overtaxed network to its limits.

Perhaps it's not surprising then that like the spark that ignites a raging bushfire, the onset of the crisis that struck on August 10 was a relatively minor event. A single transmission line in western Oregon, just north of Portland, sagged a little too far and struck a tree that had been left untrimmed a little too long, causing it to flash over. Not an uncommon occurrence really, and the operators at the Bonneville Power Authority who were instantly notified of this unwelcome but far from disastrous failure scarcely shifted in their seats. What happened next, however, was frighteningly swift and totally unexpected.

The line that failed—the Keeler-Allston line—was one of a set of parallel cables that shipped power from Seattle down to Portland, and the automatic coping mechanism for such a failure was to transfer the load to the other lines in the set. Unfortunately, these lines were also carrying loads close to their limits, and the extra burden proved too much. One by one, the dominoes began to fall. First was the adjacent Pearl-Keeler line, forced out of service because of a transformer outage. Then, about five minutes later, the St. Johns-Merwin line tripped due to a relay malfunction, the successive failures forcing large volumes of power east and then west across the Cascade Mountains and dragging the system to the brink of dangerous, large-amplitude voltage oscillations.

When power lines are heavily loaded, they heat up and stretch. By August, the trees had had all summer to grow, and by almost four o'clock in the afternoon, when the final hammer blow came, even a lightly loaded line would have been sagging in the hot sun. The hopelessly overtaxed Ross-Lexington line stretched too far and, like Keeler-Allston two hours before, hit one of the ubiquitous trees. This final disturbance proved too much for the nearby McNary generators, and their protective relays progressively tripped all thirteen of them, driving the system beyond the range of any contingency it had been designed to cope with. The incipient voltage oscillations commenced, and seventy seconds later all three strands of the California-Oregon intertie—the bottleneck through which all power passes up and down the West Coast—relayed out of service.

One of the fundamental rules about electric power is that it is extremely hard to store. You can power your cell phone or laptop with a battery for a few hours, but no one has yet developed the technology to build batteries that can power cities. As a result, power has to be generated *when* it is needed, and shipped instantly to *where* it is needed. The flip side of this rule is that once generated, it has to *go* somewhere, and that is precisely where the power that had been flowing to northern California had to go—somewhere. Cut off from California by the severed intertie, it surged east from Washington and then south, sweeping like a tidal wave through Idaho, Utah, Colorado, Arizona, New Mexico, Nevada, and southern California, tripping hundreds of lines and generators, fracturing the western system into four isolated islands, and interrupting service to 7.5 million people. That night, the skyline of San Francisco was dark. Mercifully there were no riots, which possibly says something about San Franciscans versus New Yorkers. But in the course of the cascade, 175 generating units had tripped out of service, and some—the nuclear reactors—required several days to restart, contributing to an estimated total cost of some $2 billion.

How did it happen? Well, in one sense we know exactly how it happened. Engineers from Bonneville Power and the coordinating council immediately went to work, producing a detailed disturbance report as early as mid-October. The basic problem was that too many people were asking too much from too little. Aside from that, the report blamed a number of factors, including sloppy maintenance and insufficient attention paid to warning signs. Bad luck had also come into play. Some units that might have buffered the system were either out of service for maintenance or shut down because of environmental regulations constraining hydroelectric spills in salmon-bearing rivers. Finally, the report pointed to an inadequate understanding of the interdependencies present in the system.

It is this last comment, slipped innocuously between the precisely identified and easily grasped complaints, that we should focus on, because it raises the question, What is it about the *system* that enabled the failure to occur? And in this sense of the problem, we have no idea at all. The trouble with systems like the power grid is that they are built up of many components whose individual behavior is reasonably well understood (the physics of power generation is nineteenth-century stuff) but whose collective behavior, like that of football crowds and stock market investors, can be sometimes orderly and sometimes chaotic, confusing, and even destructive. The cascading failure that struck the West on August 10, 1996, was not a sequence of independent random events that simply aggregated to the point of a crisis. Rather, the initial failure made subsequent failures more likely, and once they occurred, that made further failures more likely still, and so on.

It's one thing to say this, however, and it's quite another to understand *precisely* how certain failures under certain conditions are likely to be benign and how other failures or other conditions are recipes for disaster. One needs to think about the consequences of not just single failures but also combinations of failures, and this makes the problem truly hard. But it gets worse. Perhaps the most perturbing aspect of cascading failures, and one that was starkly exhibited by the August 10 outage, is that by installing protective relays on the power generators—by reducing, in effect, the possibility that individual elements of the system would suffer serious damage—the designers had inadvertently made the system *as a whole* more likely to suffer precisely the kind of global meltdown that occurred.

EMERGENCE

How ARE WE TO UNDERSTAND SUCH PROBLEMS? IN FACT, WHAT is it about complex connected systems that makes them so hard to understand in the first place? How is it that assembling a large collection of components into a *system* results in something altogether different from just a disassociated collection of components? How do populations of fireflies flashing, crickets chirping, or pacemaker cells beating manage to synchronize their rhythms without the aid of a central conductor? How do small outbreaks of disease become epidemics, or new ideas become crazes? How do wild speculative bubbles emerge out of the investment strategies of otherwise sensible individuals, and when they burst, how does their damage spread throughout a financial system? How vulnerable are large infrastructure networks like the power grid or the Internet to random failures or even deliberate attack? How do norms and conventions evolve and sustain themselves in human societies, and how can they be upset and even replaced? How can we locate individuals, resources, or answers in a world of overwhelming complexity, without having access to centralized repositories of information? And how do entire business firms innovate and adapt successfully when no one individual has enough information to solve or even fully understand the problems that the firm faces?

As different as these questions appear, they are all versions of the same question: *How does individual behavior aggregate to collective behavior?* As simply as it can be asked, this is one of the most fundamental and pervasive questions in all of science. A human brain, for example, is in one sense a trillion neurons connected together in a big electrochemical lump. But to all of us who have one, a brain is clearly much more, exhibiting properties like consciousness, memory, and personality, whose nature cannot be explained simply in terms of aggregations of neurons.

As the Nobel laureate Phillip Anderson explained in his famous 1971 paper, "More Is Different," physics has been reasonably successful in classifying the fundamental particles, and in describing their individual behavior and interactions, up to the scale of single atoms. But throw a bunch of atoms together, and suddenly the story is entirely different. That's why chemistry is a science of its own, not just a branch of physics. Moving farther up the chain of organization, molecular biology cannot be reduced simply to organic chemistry, and medical science is much more than the direct application of the biology of molecules. At a higher level still—that of interacting organisms—we encounter now a host of disciplines, from ecology and epidemiology, to sociology and economics, each of which comes with its own rules and principles that are not reducible to a mere knowledge of psychology and biology.

After hundreds of years of denial, modern science has finally come to terms with this way of seeing the world. The dream of Pierre Laplace, the great nineteenth-century French mathematician—that the universe could be understood in its entirety through reduction to the physics of fundamental particles, churned through a sufficiently powerful computer—has spent the better part of the last century staggering around the scientific stage like a mortally wounded Shakespearean actor, uttering its final soliloquy before finally collapsing altogether. It's just not quite clear what stands in its place. On the one hand, the idea that lumping a bunch of things together will somehow produce something other than just a bunch of things seems blindingly obvious. On the other hand, the realization that we have made so little progress on the matter should give you some idea of how hard it is.

What makes the problem hard, and what makes complex systems complex, is that the parts making up the whole don't sum up in any simple fashion. Rather they interact with each other, and in interacting, even quite simple components can generate bewildering behavior. The recent sequencing of the human genome revealed that the basic code of all human life consists of only about thirty thousand genes—many fewer than anyone had guessed. So whence comes all the complexity of human biology? Clearly it is not from the complexity of the individual elements of the genome, which could scarcely be any simpler, nor does it come from their number, which is barely any greater than it is for the humblest of organisms. Rather it derives from the simple fact that genetic traits are rarely expressed by single genes. Although genes, like people, exist as identifiably individual units, they *function* by interacting, and the corresponding patterns of interactions can display almost unlimited complexity.

What then of human systems? If the interactions of mere genes can confound the best minds in biology, what hope do we have of understanding combinations of far more complex components like people in a society or firms in an economy? Surely the interactions of entities that are themselves complex would produce complexity of a truly intractable kind. Fortunately, as capricious, confusing, and unpredictable as individual humans typically are, when many of them get together, it is sometimes the case that we can understand the basic organizing principles while ignoring many of the complicated details. This is the flip side of complex systems. While knowing the rules that govern the behavior of individuals does not necessarily help us to predict the behavior of the mob, we *may* be able to predict the very same mob behavior without knowing very much at all about the unique personalities and characteristics of the individuals that make it up.

An apocryphal story that illustrates this last point is the following. A few years ago in the United Kingdom, power utility engineers were puzzled by peculiar synchronized surges of demand that were draining many parts of the national grid simultaneously and seriously overtaxing its production capacity, albeit only for minutes at a time. They finally figured it out when they realized that the worst of all such surges occurred during the annual soccer championship, during which the entire country was glued to their television sets. At halftime, a nation of soccer fans got up from their couches, virtually simultaneously, and put the kettle on for a cup of tea. Although individually, the British are as complicated as anyone, you don't need to know much about each of them to figure out the surges in power demand—just that they like soccer and tea. In this case a quite simple representation of the individuals works just fine.

Sometimes, therefore, the interactions of individuals in a large system can generate greater complexity than the individuals themselves display, and sometimes much less. Either way, the particular manner in which they interact can have profound consequences for the sorts of new phenomena—from population genetics to global synchrony to political revolutions—that can *emerge* at the level of groups, systems, and populations. As with the cascade in the power grid, however, it is one thing to state this and quite another matter altogether to understand it precisely. In particular, what is it about the patterns of interactions between individuals in a large system that we should pay attention to? No one has the answer yet, but in recent years a growing group of researchers has been chasing a promising new lead. And out of this work, which in itself builds on decades of theory and experiment in every field from physics to sociology, is coming a new science, the science of networks.

NETWORKS

In A WAY, NOTHING COULD BE SIMPLER THAN A NETWORK. STRIPPED to its bare bones, a network is nothing more than a collection of objects connected to each other in some fashion. On the other hand, the sheer generality of the term *network* makes it slippery to pin down precisely, and this is one reason why a science of networks is an important undertaking. We could be talking about people in a network of friendships or a large organization, routers along the backbone of the Internet or neurons firing in the brain. All these systems are networks, but all are completely distinct in one sense or another. By constructing a language for talking about networks that is precise enough to describe not only what a network is but also what kinds of different networks there are in the world, the science of networks is lending the concept real analytical power.

But why is this new? As any mathematician can tell you, networks have been studied as mathematical objects called *graphs* ever since 1736, when Leonhard Euler, one of the greatest mathematicians ever, realized that the problem of taking a stroll across all seven bridges in the Prussian city of Königsberg without crossing the same bridge twice could be formulated as a graph. (He proved, incidentally, that it couldn't be done, and that was the first theorem in graph theory.) Since Euler, the theory of graphs has grown steadily to become a major branch of mathematics and has spilled over into sociology and anthropology, engineering and computer science, physics, biology, and economics. Each field therefore has its own version of a theory of networks, just as each has its own way of aggregating individual to collective behavior. So why is there anything fundamental left to figure out?

The crux of the matter is that in the past, networks have been viewed as objects of *pure structure* whose properties are *fixed in time*. Neither of these assumptions could be further from the truth. First, real networks represent populations of individual components that are actually *doing something*—generating power, sending data, or even making decisions. Although the structure of the relationships between a network's components is interesting, it is *important* principally because it affects either their individual behavior or the behavior of the system as a whole. Second, networks are dynamic objects not just because things happen in networked systems, but because the networks themselves are evolving and changing in time, driven by the activities or decisions of those very components. In the connected age, therefore, *what happens and how it happens depend on the network*. And the network in turn depends on what has happened previously. It is this view of a network—as an integral part of a continuously evolving and self-constituting system—that is truly new about the science of networks.

Understanding networks in this more universal fashion, however, is an extraordinarily difficult task. Not only is it inherently complicated, but also it requires different kinds of specialized knowledge that are usually segregated according to academic specialty and even discipline. Physicists and mathematicians have at their disposal mind-blowing analytical and computational skills, but typically they don't spend a whole lot of time thinking about individual behavior, institutional incentives, or cultural norms. Sociologists, psychologists, and anthropologists, on the other hand, do. And in the past half century or so they have thought more deeply and carefully about the relationship between networks and society than anyone else—thinking that is now turning out to be relevant to a surprising range of problems from biology to engineering. But, lacking the glittering tools of their cousins in the mathematical sciences, the social scientists have been more or less stalled on their grand project for decades.

If it is to succeed, therefore, the new science of networks must bring together from all the disciplines the relevant ideas and the people who understand them. The science of networks must become, in short, a manifestation of its own subject matter, a network of scientists collectively solving problems that cannot be solved by any single individual or even any single discipline. It's a daunting task, made all the more awkward by the long-standing barriers separating scientists themselves. The languages in the various disciplines are very different, and we scientists often have difficulty understanding one another. Our approaches are different too, so each of us has to learn not only how the others speak but also how they think. But it *is* happening, and the past few years have seen an explosion in research and interest around the world in search of a new paradigm with which to describe, explain, and ultimately understand the connected age. We are not there yet, not by a long shot, but as the story in the pages ahead relates, we are making some exciting progress.

SYNCHRONY

My PART IN THIS STORY BEGAN, AS MANY STORIES DO, MORE OR less by accident, in a small town in upstate New York called Ithaca. And a place named after the mythical home of Odysseus is, I suppose, as good a place to begin a story as any. Back then, however, the only Odysseus I knew was a small cricket, who along with his brothers Prometheus and Hercules was part of an experiment I was running as a graduate student at Cornell University with my adviser, Steven Strogatz. Steve is a mathematician, but pretty early on in his career he started to become much more interested in the applications of mathematics to problems in biology, physics, and even sociology than in the math itself. Even as an undergraduate at Princeton University in the early 1980s, he couldn't keep himself from dragging math into his other studies. For his sociology requirement, Steve convinced his instructor that he should be allowed to do a math project rather than write a term paper. The instructor agreed but remained somewhat puzzled. What sort of math could you do for introductory sociology? Steve chose to study romantic relationships, formulating and solving a simple set of equations that described the interaction of two lovers, Romeo and Juliet. As unlikely as it sounds, over fifteen years later at a conference in Milan, I was accosted by an Italian scientist who was so excited by Steve's work, he was trying to apply it to the plots of Italian romance films.

Steve went on to win a Marshall scholarship and undertook the formidable mathematics tripos at Cambridge University, immortalized by the great G. H. Hardy in his memoir *A Mathematician's Apology*. He didn't like it much and soon found himself longing for his homeland and a research problem he could really sink his teeth into. Fortunately, he found Arthur Winfree, a mathematical biologist who pioneered the study of biological *oscillators*—rhythmically cycling entities like neurons firing in the brain, pacemaker cells beating in the heart, and fireflies flashing in trees. Winfree (who, coincidentally, was an undergraduate at Cornell) quickly set Steve en route to his own career, collaborating on a project to analyze the structure of scroll waves in the human heart. Scroll waves are the waves of electricity that commence in the pacemaker cells and spread throughout the heart muscle, stimulating and regulating its beat. They are important to understand because sometimes they stop or lose their coherence, a potentially devastating event commonly called arrhythmia. No one has done more to understand the dynamics of the heart than Art Winfree, and although Steve quickly moved away from that particular project, he remained entranced by oscillations and cycles, particularly in biological systems.

For his Ph.D. dissertation at Harvard University, Strogatz performed an exhaustive (and exhausting!) data analysis of the human sleep-wake cycle, attempting to crack the code of circadian rhythms that lead, among other things, to the experience of jet lag when we travel across time zones. He never succeeded, and the experience spurred him to think about somewhat simpler biological cycles in more mathematical terms, which is when he started working with Rene Mirollo, a mathematician at Boston University. Inspired by the work of the Japanese physicist Yoshiko Kuramoto (who in turn had been inspired by none other than Art Winfree), Strogatz and Mirollo wrote a number of influential papers on the mathematical properties of a particularly simple class of oscillators called, appropriately enough, *Kuramoto oscillators*. The essential problem in which they and many others were interested was that of synchronization: Under what conditions will a population of oscillators start to oscillate in synchrony? Like so many of the questions in this story, this one is essentially about the emergence of some global behavior from the interactions of many individuals. Synchronization of oscillators just happens to be a particularly simple and well-defined version of emergence, and so is one part of a generally murky subject that has been understood reasonably well.

Picture a group of runners, completing laps of a circular track (Figure 1.1). Regardless of the circumstances—a collection of Sunday afternoon joggers running at the local track or prize athletes competing in an Olympic final—the members of the group will tend to vary in their natural ability. That is, if they were running individually, some would set lap times faster, and some slower, than the average. You might expect, then, that their natural variations would cause them to spread out uniformly around the track, with the very fast ones occasionally lapping the very slow ones. But we know from experience that this is not always the case. It *is* the case when the runners aren't paying any attention to each other, so perhaps the Sunday afternoon joggers

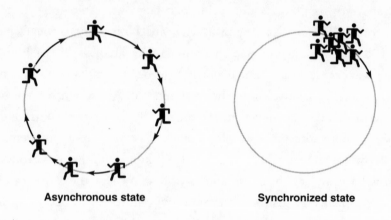

Asynchronous state   **Synchronized state**

Figure 1.1 Coupled oscillators can be visualized as
runners completing laps around a circular track. When the oscillators are
strongly coupled, they will synchronize (right); otherwise, the system
will tend toward an asynchronous state (left).

will remain quite spread out, like on the left side of Figure 1.1. But in an Olympic event, where each runner has a great incentive to stay within striking distance of the leader (and the leader has an equal incentive not to burn out prematurely), the runners pay great attention to each other, and the result is the pack (as in the right side of the figure).

In oscillator terms, the pack represents a *synchronized state,* and whether or not the system synchronizes depends both on the distribution of *intrinsic frequencies* (their individual lap times) and on the *coupling strength* (how much attention they pay to one another). If they all have the same ability and they start together, they will remain synchronized regardless of their coupling. If their distribution of abilities is great, such as in the final sprint of a ten-thousand-meter race, then no matter how much they want to stay together, the pack will disintegrate and synchrony will be lost. As simple a model as this is, it turns out to be a nice representation of many interesting systems in biology, ranging from pacemaker cells to fireflies flashing to crickets chirping. Strogatz also studied the mathematics of physical systems, like arrays of superconducting Josephson junctions—extremely fast switches that might one day form the basis of a new generation of computers.

By the time he arrived at Cornell in 1994, Steve had become a leading player in the field of coupled oscillator dynamics, had written the definitive introductory textbook on nonlinear dynamics and chaos, and had achieved his adolescent dream of tenure at a leading research university. He had won teaching prizes and research prizes, had studied and worked at some of the best universities in world—Princeton, Cambridge, Harvard, and MIT—and by his mid-thirties had a gold-plated résumé. He was, however, bored. Not unhappily so, but for ten years he had been doing roughly the same sort of thing. He felt he had mastered that particular corner of the academic universe as much as he was ever likely to, and he was ready to start exploring again. But where?

My first interaction with Steve came when he was still at MIT and I was a first-year graduate student at Cornell. Like many grad students, I had harbored dreamy visions of life in a research university and was pretty disillusioned by the difficult and often dull reality. I had decided that any place would be better than Cornell. This guy Strogatz had recently given a talk in my department—the first such talk that I felt I had actually understood—so I called him to see if he could take on a new research assistant. He replied that, in fact, he was moving to Cornell, to my very department (his talk, it turned out, had been a part of his job interview). So I stayed where I was.

As it happened, the procedure for graduate students in my department was that at the end of the first year, they undertook a qualifying exam—the Q *exam*—that is designed to test their working knowledge of basically everything they should have learned during their undergraduate education and first year of grad school. The exam is an oral one, so each student would walk into a room full of professors, and get peppered with questions that they were to answer on the blackboard. If a student passed, he or she was allowed to continue on for a Ph.D. And if they failed? Well, they weren't supposed to fail. Naturally enough then, it's a somewhat terrifying experience (although most of the terror is in the anticipation), and as luck would have it, the one question that Strogatz asked me was one I hadn't studied for—at all. After tap-dancing around the blackboard for a few minutes, during which time my lack of preparation became entirely apparent, I was mercifully spared any further humiliation, and we moved on to the next question. Fortunately, the rest of the exam went alright and, much to my relief, I passed (we all did). A week or two later, after yet another incomprehensible department seminar, Steve approached me and, much to my surprise, suggested that we talk about working together.

A slightly bored master and a mostly lost student hardly sounds like a perfect match, but it was. Over the next couple of years, we dithered through a variety of projects, spending as much time discussing philosophy as math—not the existential kind, mind you, but the practical. Which questions were interesting, and which were just difficult? Whose work did we admire and why? How much did technical mastery matter compared with creativity and daring? And how much time should one spend learning about other people's work before launching into unfamiliar territory? What did it mean, in other words, to do interesting science? As I suspect is the case with most philosophy, the answers—inasmuch as we ever came up with any—were less important than the process of thinking through the questions, and that process deeply influenced our subsequent work. The dithering, it turned out, had not only allowed us to become friends and given me a chance to finish my courses; it also freed us from the focus of a single, well-defined project for long enough to think about what we really *wanted* to do, rather than just what we thought we *could* do. And that, as the poem goes, has made all the difference.

THE ROAD LESS TRAVELED

AT THE TIME WE STUMBLED ONTO OUR EVENTUAL PROJECT, WE were studying, of all things, crickets. It sounds silly, but because this particular species of cricket—the snowy tree cricket—chirps in such a regular fashion and because (unlike pacemaker cells or neurons) it is such a well-behaved experimental subject, it is virtually an ideal specimen of biological oscillator. We were trying to test a deep mathematical hypothesis, originally proposed by Winfree, that only certain types of oscillators can synchronize. Since snowy tree crickets are extremely good at synchronizing, it seemed a natural step to determine *experimentally* what kind of oscillators they were and therefore whether or not the theoretical predictions were true.

Crickets, not surprisingly, are also of interest to biologists, and because chirping is intimately related to mating and reproductive success, the mechanisms leading to global synchrony are also important biological issues. As a result, Steve and I were working with an entomologist, Tim Forrest, with whom I had spent several late-summer evenings scrambling around the trees of Cornell's enormous campus searching for specimens, including the aforementioned Odysseus. Having assembled our small troop, we then proceeded to isolate each one in a soundproof chamber and chirp at it with the aid of a computer that Tim had rigged up to a speaker-microphone system. Recording their responses to the computer's precisely timed stimuli, we were able to characterize how much a cricket advances or delays its next chirp depending on the point in its natural cycle at which it hears the other "cricket," which in this case was our computer (crickets, apparently, are easily fooled).

That, however, was the easy part. The situation we had contrived was incredibly artificial—a single cricket in a soundproof chamber chirping in isolation with the occasional prod from a computer that wasn't even listening in turn. In the real world, this would never happen. Not only do crickets listen and respond to each other, but also in any single bush or tree there would typically be many crickets, all doing the same thing. The question in my mind was, *Who was listening to whom?* Surely there was no master cricket from which all others took their cues. But if not, then how did they manage to synchronize so well? Does each cricket listen to every other cricket? Or only one other? Or maybe just a few? What structure, if any, is there in the population, and does it in fact matter?

Back then I wasn't accustomed to seeing networks everywhere I looked, but even so it occurred to me that the pattern of interactions— the *coupling topology* in the parlance of oscillator theory—could be thought of as some kind of network. It also occurred to me that whatever structure that network exhibited might influence the ability of a population to synchronize, and therefore might be important to understand as an entity in its own right. Thinking like a typical graduate student, I assumed that the coupling topology question was an obvious one, and therefore the answer must have been worked out long ago—all I needed to do was look it up. Instead of the answer, however, I only found more questions. Not only was the relationship between network structure and oscillator synchronization almost completely unexplored, but also no one really seemed to have thought much about the relationship between networks and *any* kind of dynamics. Even the more basic question of what sorts of networks existed in the real world seemed to have escaped attention—at least from mathematicians. It began to dawn on me that I had stumbled across what every graduate student hopes to find but rarely does—a genuine hole in science, an undiscovered door that is open just a crack, a chance to explore the world in a new way.

Right about this time, I remembered something my Dad had mentioned to me over a year earlier, during a phone call on a Friday evening. For a reason that we have both forgotten, he had asked me if I had ever heard of the notion that no one was ever more than "six degrees from the president." That is, you know someone, who knows someone, who knows someone, . . . who knows the president of the United States. I hadn't heard of it, and I remember being on a Greyhound bus somewhere between Ithaca and Rochester trying to figure out how it might be possible. I hadn't made any progress that day or since, but I did remember thinking about the problem as some kind of network of relationships between individuals. Each person has a circle of acquaintances—network neighbors—who in turn have acquaintances, and so on, forming a global interlocking pattern of friendship, business, family, and community ties through which paths could be traced between any random person and any other. It occurred to me now that the length of those paths might have something to do with the way that influences—whether they be diseases, rumors, ideas, or social unrest—propagate through a human population. And if the same *six degrees* property also happened to be true of nonhuman networks, like biological oscillators for instance, then it might be important for understanding phenomena like synchronization.

All of a sudden, the quaint urban myth that my father had related to me seemed terribly important, and I was determined to get to the bottom of it. Several years later, we are still getting there. The hole, it turns out, is pretty deep, and it will be many more years before it is fully explored and mapped out. But already we have made some quite tidy progress. We have also learned a good deal about this six degrees problem, which is not an urban myth at all but a sociological research project with a storied history.

THE SMALL-WORLD PROBLEM

IN 1967, THE SOCIAL PSYCHOLOGIST STANLEY MILGRAM PERFORMED a remarkable experiment. Milgram was interested in an unresolved hypothesis circulating in the sociological community of the day. The hypothesis was that the world, viewed as an enormous network of social acquaintances, was in a certain sense "small"; that is, any one person in the world could be reached through a network of friends in a only a few steps. It was called the *small-world problem,* after the cocktail party banter in which two strangers discover that they have a mutual acquaintance and remind each other what a "small world" it is (this happens to me a *lot*).

Actually, the cocktail party observation is not really the same as the small-world problem that Milgram was studying. Only a small fraction of people in the world can possibly have mutual acquaintances, and the fact that we seem to run into them with surprising regularity has more to do with our tendency to pay attention to the things that surprise us (and thus overestimate their frequency) than it does with social networks. What Milgram wanted to show was that even when I *don't* know someone who knows you (in other words, all those times we meet people and don't end up saying "what a small world"), I still know someone, who knows someone, who knows someone who does know you. Milgram's question was, How many *someones* are in the chain?

To answer that question, Milgram devised an innovative message-passing technique that is still known as the *small-world method.* He gave letters to a few hundred, randomly selected people from Boston and Omaha, Nebraska. The letters were to be sent to a single *target* person, a stockbroker from Sharon, Massachusetts, who worked in Boston. But the letters came with an unusual rule. Recipients could only send their letter on to somebody whom they knew on a first-name basis. Of course, if the recipients knew the target person, then they could send it to him directly. But if they didn't, and it was extremely unlikely that they would, they were to send it to someone they did know who they thought was somehow closer to the target.

Milgram was teaching at Harvard at the time, so naturally he regarded the greater Boston area as the center of the universe. And what else could be farther from it than Nebraska? Not only geographically but also socially, the Midwest seemed impossibly distant. When Milgram asked people how many steps it would take to get a letter from one place to the other, they typically estimated it in the hundreds. The result was more like six—a result that was so surprising at the time, it led to the phrase "six degrees of separation," after which John Guare named his 1990 play and which has spawned a number of parlor games, not to mention an infinite number of conversations at cocktail parties.

But why exactly was Milgram's finding so surprising? If you're mathematically inclined, you might do the following thought experiment, perhaps even drawing a picture like the one in Figure 1.2.

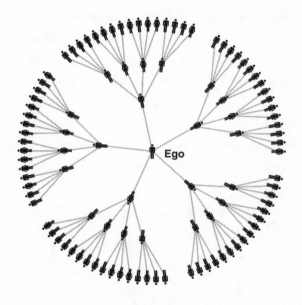

Figure 1.2. A pure branching network. Ego knows only 5 people, but within two degrees of separation, ego can reach 25; within three degrees, 105; and so on.

Imagine I have a hundred friends, each one of which has a hundred friends also. So at one degree of separation I connect to one hundred people, and within two degrees I can reach one hundred times one hundred, which is ten thousand people. By three degrees, I am up to almost one million; by four, nearly a hundred million; and in five degrees, about nine billion people. In other words, if everyone in the world has only one hundred friends, then within six steps, I can easily connect myself to the population of the entire planet. So maybe it's obvious that the world is small.

If you are at all socially inclined, however, you have already spotted the fatal flaw in this reasoning. A hundred people is a lot to think about, so think about your ten best friends and ask yourself who their ten best friends are. Chances are you will come up with many of the same people. This observation turns out to be an almost universal feature not just of social networks but of networks in general. They display what we call *clustering*, which is really just to say that most people's friends are also to some extent friends of each other. In fact, social networks are much more like Figure 1.3. We tend not so much to have friends as we do groups of friends, each of which is like a little cluster based on shared experience, location, or interests, joined to each other by the overlaps created when individuals in one group also belong to other groups. This characteristic of networks is particularly relevant to the small-world problem, because clustering breeds redundancy. In particular, the more

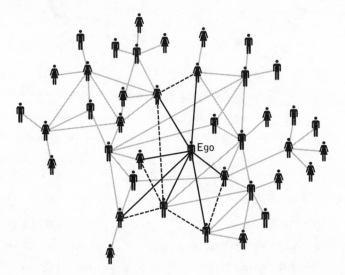

Figure 1.3. Real social networks exhibit clustering, the tendency of two individuals who share a mutual friend to be friends themselves. Here, Ego has six friends, each of whom is friends with at least one other.

your friends know each other, the less use they are to you in getting a message to someone you don't know.

The paradox of social networks that Milgram's experiment illuminated is that on the one hand, the world is highly clustered—many of my friends are also friends of each other. Yet on the other hand, we can still manage to reach anyone at all in an average of only a few steps. Although Milgram's small-world hypothesis has stood largely unchallenged for over three decades, it remains surprising to this day. As Ouisa says in Guare's play, "everybody on this planet is separated by only six other people. Six degrees of separation. Between us and everybody else on this planet. The president of the United States. A gondolier in Venice. . . . It's not just the big names. It's *anyone*. A native in a rain forest. A Tierra del Fuegan. An Eskimo. I am bound to everyone on this planet by a trail of six people. It's a profound thought."

And it is a profound thought. If we are thinking only about a certain subgroup of people, people with whom we have something obvious in common, then again we might think the result is hardly surprising. For example, I teach at a university, and because the university world consists of a relatively small number of people, many of whom have a good deal in common, it's relatively easy for me to imagine how I might pass a message through a sequence of colleagues to any other university professor anywhere in the world. Similar reasoning might convince you that I could get a message to most college-educated professionals in the New York area. But this is not really the small-world phenomenon—it's more like a small-group phenomenon. The claim of the small-world phenomenon is much stronger than this. It states that I can get a message to *anyone*, even if they have absolutely nothing in common with me at all. Now it seems much less obvious, if only because human society is so deeply cleaved along the fault lines of race, class, religion, and nationality.

For thirty years or more, while the small-world phenomenon grew from a sociological conjecture into a piece of pop-culture folklore, the actual nature of the world remained in question, and the paradox at its heart—that apparently distant people can actually be very close—remained just that, a paradox. But the last few years have seen a burst of theoretical and empirical work, mostly from outside of sociology, that not only has helped to resolve the small-world phenomenon but also suggests it is far more general than anyone ever realized. This rediscovery, in a sense, of the small-world phenomenon, known so long only to sociologists, has led to an even broader set of questions about networks that are relevant to a whole host of applications in science, business, and everyday life.

As is so often the case in science (and even everyday problem solving), the idea that broke the stalemate was found by coming at an old issue from a new direction. Instead of asking, "How small is our world?" one could ask, "What would it take for any world, not just ours, but any world at all, to be small?" In other words, rather than going out into the world and measuring it in great detail, we want to construct a mathematical model of a social network, in place of the real thing, and bring to bear on our model the power of mathematics and computers. The networks we will actually be dealing with can be represented in almost comical simplicity by dots on a piece of paper, with lines connecting them. In mathematics, such objects are called graphs, and as we saw earlier, the study of graphs is a centuries-old subject about which an enormous amount is known. And that's the point really. Although in making such a drastic simplification, we inevitably miss features of the world that we ultimately care about, we can tap into a wealth of knowledge and techniques that will enable us to address a set of very general questions about networks that we might never have been able to answer had we gotten bogged down in all the messy details.

CHAPTER TWO

The Origins of a "New" Science

• • • • • •

THE THEORY OF RANDOM GRAPHS

ABOUT FORTY YEARS AGO, THE MATHEMATICIAN PAUL ERDŐS took a particularly simple approach to the study of communication networks. Erdős was the kind of unusual figure who makes other oddballs look positively vanilla. Born in Budapest on March 26, 1913, Erdős lived with his mother until he was twenty-one, then spent the rest of his remarkable life living out of two battered suitcases. Never staying anywhere for long and never holding a permanent job, Erdős relied on the hospitality of his devoted colleagues, who were more than happy to oblige him in return for the company of his lightning-fast, ever questioning mind. He regarded himself famously as a mechanism for turning coffee into theorems—not that he ever actually learned to make coffee or to do many other everyday tasks, like cooking and driving, that lesser mortals generally find quite straightforward. When it came to mathematics, however, he was an absolute juggernaut, publishing nearly fifteen hundred papers in his lifetime (and even a few after), more than any mathematician in history, bar possibly the great Euler.

He also invented, with his collaborator Alfred Rényi, the formal theory of random graphs. A *random graph* is, as the name might suggest, a network of nodes connected by links in a purely random fashion. To use an analogy of the biologist Stuart Kauffman, imagine throwing a boxload of buttons onto the floor, then choosing pairs of buttons at random and tying them together with appropriate-length threads of string (Figure 2.1). If we have a very large floor, a very large box of but-

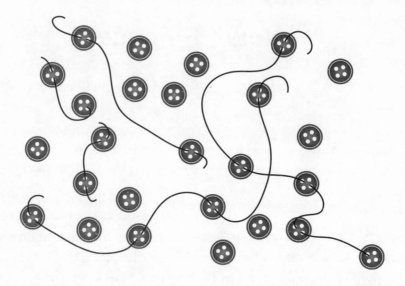

Figure 2.1. A random graph imagined as a collection of buttons tied by strings. Pairs of nodes (buttons) are connected at random by links or ties.

tons, and plenty of spare time, what would such networks end up looking like? In particular, what features could we *prove* that all such networks must have? The word *prove* is what makes random graph theory hard. Very hard. It's not enough simply to try a few examples and see what happens. One needs to consider what can happen and what cannot happen in every conceivable circumstance, and what sorts of conditions must hold in order to be sure. Fortunately, Erdős was the master of proof, and one particularly deep result that he and Rényi proved is the following.

Returning to the button metaphor, imagine attaching a number of threads to buttons, any number you like, and then picking a button at random, counting all the other buttons that come off the floor with it. All of these secondary buttons are part of the chosen button's *connected component*. If we repeat the exercise by picking up one of the remaining buttons, we will find another connected component, and we can keep going in this manner until all the buttons have been removed from the floor. How big will the largest of these components be? That will depend on how many threads you have attached. But how *exactly* does it depend?

If you have a thousand buttons and you have attached only one thread, the largest component will contain only two buttons, which, as a fraction of the whole network, is close to zero. If at the other extreme you have attached every button to every single other button, then equally obviously the largest component will include all one thousand buttons, or the whole network. But what happens for all possible cases in between? Figure 2.2 is a plot of the fraction of the network, or random graph, occupied by its largest connected component versus the number of links present. As expected, when we have very few links, nothing is connected to anything. Because we have added the threads purely at random, we will almost always be connecting one isolated button to another, and even if by chance one of them already has a thread on it, that thread probably only leads to a small number of other buttons.

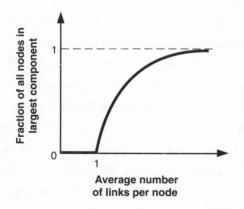

Figure 2.2.
Connectivity of a random graph. The fraction of the nodes connected in a single component changes suddenly when the average number of links per node exceeds one.

But then something odd happens. When we have added enough threads such that each button has *on average* one thread attached to it, the fraction of the graph occupied by the largest component jumps suddenly and rapidly from almost zero to nearly one. In the language of physics, this rapid change is called a *phase transition* because it transitions between a disconnected *phase* and a connected one, and the point at which this starts to happen (where the line first kicks up in Figure 2.2) is called the *critical point*. As we will see, phase transitions of one sort or another occur in many complex systems and have been used to explain phenomena as diverse as the onset of magnetization, the explosion of disease epidemics, and the propagation of cultural fads. In this particular case, the phase transition is driven by the addition of a small number of links right near the critical point that have the effect of connecting many very small clusters into a single *giant component*, which then proceeds to swallow up all the other nodes until everything is connected. The existence and nature of this phase transition are what Erdős and Rényi explained back in 1959.

Why should we care about this? Simply put, if two nodes are not part of the same component, they cannot communicate, interact, or otherwise affect each other. They might as well be in different systems, in that the behavior of one cannot possibly have anything to do with the behavior of the other. So the presence of a giant component means that whatever happens at one location in the network has the potential to affect any other location. Its absence, by contrast, implies that local events are only felt locally. Erdős and Rényi motivated their initial work by thinking about communication networks. They asked how many links would have to be laid down between a group of devices before one picked at random would be able to communicate with the bulk of the system. The line between isolation and connectedness is thus an important threshold for the flow of information, disease, money, innovations, fads, social norms, and pretty much everything else that we care about in modern society. That global connectivity should arrive not incrementally but in a sudden, dramatic jump tells us something deep and mysterious about the world—at least if we believe that random graphs tell us anything about the world.

And that, of course, is the problem. As sophisticated as the theory of random graphs is (and it is bewilderingly sophisticated), almost everything we know about real networks, from social networks to neural networks, suggests that they are not random, or at least nothing like the random graphs of Erdős and Rényi. Why? Well, imagine that you really did pick your friends at random from the global population of over six billion. You would be much more likely to be friends with someone on another continent than someone from your hometown, workplace, or school. Even in a world of global travel and electronic communications, this is an absurd notion. But continuing with it for a bit, even if you had, say, one thousand friends, and each of them also had a thousand friends, the chance that any of your friends would know each other would be roughly one in six million! Yet we know from everyday experience that our friends *do* tend to know one another, so random graphs cannot be a good representation of the real social world. Unfortunately, as we will see later, as soon as we depart from the highly idealized assumptions of pure randomness that the graph theorists rely on, it becomes extremely hard to prove anything at all. Nevertheless, if we would like to understand the properties and behavior of real-world networks, the issue of nonrandom structure is one that eventually has to be faced.

SOCIAL NETWORKS

IT IS ONLY SLIGHTLY UNFAIR TO CHARACTERIZE SOCIOLOGY AS A discipline that attempts to explain human behavior without the humans. Whereas psychology is very much concerned with understanding what people do in terms of their individual characteristics, experiences, and even physiology, sociology tends to regard human action, or *agency*, as constrained, even determined, by the roles that people play within the political, economic, and cultural institutions that define their social environment. Or as Marx put it, "Men make their own history, but . . . they do not make it under circumstances chosen by themselves." Sociology is therefore all about structure. Perhaps it's not surprising then that the theory of network analysis that grew out of sociology (and its sister discipline, anthropology) has always exhibited a strong structuralist flavor.

To grossly compress five decades of thought into a few pages, social network analysts have developed two broad strands of techniques for thinking about networks. The first strand deals with the relationship between *network structure*—the observed set of ties linking the members of a population like a firm, a school, or a political organization—and the corresponding *social structure*, according to which individuals can be differentiated by their membership in socially distinct groups or roles. A substantial array of definitions and techniques have been introduced over the years, bearing exotic names like *blockmodels, hierarchical clustering,* and *multidimensional scaling.* But all of them are essentially designed to extract information about socially distinct groups from purely relational network data, either in terms of some direct measure of "social distance" between actors or by grouping actors according to how similar their relations are to other actors in the network. Networks, according to this view, are the signature of social identity—the pattern of relations between individuals is a mapping of the underlying preferences and characteristics of the individuals themselves.

The second strand of techniques bears a more mechanistic flavor. Here the network is viewed as a conduit for the propagation of information or the exertion of influence, and an individual's place in the overall pattern of relations determines what information that person has access to or, correspondingly, whom he or she is in a position to influence. A person's social role therefore depends not only on the groups to which he or she belongs but also on his or her positions within those groups.

As with the first strand, a number of metrics have been developed to quantify individuals' network positions and to correlate their numerical values with observable differences in individual performance.

Something of an exception to both these general categories, and a precursor to some of the models that we will encounter later in the guise of the small-world problem, is the concept known as a *weak tie,* introduced by the sociologist Mark Granovetter. After completing an extensive study of two Boston communities whose attempts to mobilize against the threat of urban development had starkly different outcomes, Granovetter came to the surprising conclusion that effective social coordination does not arise from densely interlocking "strong" ties. Rather it derives from the presence of occasional weak ties between individuals who frequently didn't know each other that well or have much in common. In his seminal 1973 paper, he called this effect "the strength of weak ties," a beautiful and elegant phrase that has since entered the lexicon of sociology.

Granovetter later showed a similar correlation between weak ties and an individual's prospects of getting a job. Job hunting, it turns out, is not just a matter of having a friend to get you in the door— precisely what sort of friend they are is of great importance. Paradoxically, however, it is not your close friends who are of most use to you. Because they know many of the same people you do, and may often be exposed to similar information, they are rarely the ones who can help you leap into a new environment, no matter how much they want to. Instead it tends to be casual acquaintances who are the useful ones because they can give you information you would never otherwise have received.

Weak ties, furthermore, can be thought of as a link between individual- and group-level analysis in that they are created by individuals, but their presence affects the status and performance not just of the individuals who "own" them but of the entire group to which they belong. Correspondingly, Granovetter claimed that only by looking at structure on a group level—by observing the structure in which the individuals are embedded—would it be possible to distinguish ties as either strong or weak. Although we will later see that the relation between the local (individual) and the global (groups, communities, populations, and so on) is somewhat more subtle than Granovetter described thirty years ago, his work represents a remarkable premonition of what is now the new science of networks.

THE DYNAMICS MATTERS

THE SOCIAL NETWORK ANALYSTS' DEEP UNDERSTANDING OF structure opens the door to a whole range of questions that are essentially off-limits to pure graph theory. But social network analysis still has a major problem: *there is no dynamics.* Instead of thinking of networks as entities that evolve under the influence of social forces, network analysts have tended to treat them effectively as the frozen embodiment of those forces. And instead of regarding networks as merely the conduits through which influence propagates according to its own rules, the networks themselves were taken as a direct representation of influence. In this way of thinking, the network structure, regarded as a static set of metrics, is thought to manifest all the information about social structure that is relevant to the behavior of individuals and their ability to influence the behavior of the system. All one needs to do is collect network data and then measure the right properties, and miraculously all will be revealed.

But what should one measure? And what exactly does it reveal? The answers, it turns out, can depend a lot on precisely what sort of application one is dealing with. The spread of a disease, for example, is not necessarily the same as the spread of a financial crisis or the diffusion of a technological innovation. The structural features of the network that enable an organization to gather information efficiently may not be the same as those that enable it to process the information it already has or to recover from some unexpected catastrophe. Six degrees from the president of the United States may be a short distance or a long one, depending on what it is that you are trying to do. Or as Jon Kleinberg (whose inspiring work on the small-world problem we will encounter in chapter 5) once explained to a reporter, he (Kleinberg) has written papers with a University of California, Berkeley, scholar who once collaborated with the future CEO of Microsoft. "Sadly," Kleinberg observed, "this gives me little influence with Bill Gates."

Because purely structural and static measures of network structure can't account for whatever action is taking place on the network, the methods offer no systematic way to translate their output into meaningful statements about outcomes. By analogy, consider a school of management that claims that leadership is an entirely generic skill whose rules apply universally. The appeal of such a school is obvious—learn how to "manage" and you can manage anything from a start-up to a nonprofit to an army platoon—but in practice it doesn't work that simply. The kind of leadership required in a combat infantry unit, for example, is dramatically different from that in a government bureau, and a leader who does very well in one environment may well do very poorly in another. This is not to say that there are no common principles at all; rather the principles must be interpreted in light of what the particular organization is trying to achieve and the kinds of people who work there. The same is true of structural analysis. Without a corresponding theory of behavior—of *dynamics*—a theory of network structure is essentially uninterpretable and, therefore, of little practical use.

An important example of how a purely structural approach to networks has led many analysts into a reassuring but ultimately misleading view of the world is the case of *centrality*. One of the great mysteries of large distributed systems—from communities and organizations to brains and ecosystems—is how globally coherent activity can emerge in the absence of centralized authority or control. In systems like dictatorships and satellite pager networks that have been specifically designed and engineered to exert control, the problem of decentralized coordination is usually avoided by explicitly building in a control center. But in many systems, usually those that have developed or evolved naturally, the source of control is far from clear. Nevertheless, the intuitive appeal of centrality is such a strong one that network analysts have focused heavily on devising centrality measures either for individuals in a network or for the network as a whole.

Implicit in this approach is the assumption that networks that appear to be decentralized really are not at all. If we look carefully at the network data, it claims, even a large and complex network will reveal itself to hinge on some small subset of influential players, information brokers, and critical resources, which together form the functional center on which everyone else depends. These key players may not be obvious—they might seem to be unimportant by conventional measures of status and power—yet they are always there. And once they are identified, we are back on familiar terrain, dealing with a system that has a center. Notions of centrality have been enormously popular in the network literature, and it's easy to see why. The theory is empirical and analytical, it produces quantifiable results that are sometimes surprising (the most coherent power base in a firm turns out to be the smokers, who gather outside several times each day; the boss's assistant, not the boss, is the key information broker), yet it doesn't force us to stomach any truly difficult or counterintuitive notions. The world always has a center, information is processed and distributed by the center, and central players wield more influence than peripheral players.

But what if there just isn't any center? Or what if there are many "centers" that are not necessarily coordinated or even on the same side? What if important innovations originate not in the core of a network but in its peripheries, where the chief information brokers are too busy to watch? What if small events percolate through obscure places by happenstance and random encounters, triggering a multitude of individual decisions, each made in the absence of any grand plan, yet aggregating somehow into a momentous event unanticipated by any-one, *including the actors themselves?*

In such cases, the network centrality of individuals, or any central-ity for that matter, would tell us little or nothing about the outcome, *because the center emerges only as a consequence of the event itself.* This statement has tremendous implications for our understanding of net-works. In a multitude of systems from economics to biology, events are driven not by any preexisting center but by the interactions of equals. Remember the last time you were in a large crowd at a concert, and in the midst of chaotic applause, the entire audience suddenly started clapping in unison. Did you ever wonder how everyone manages to agree on a single beat? After all, many people naturally clap at different rates, and they don't all start at exactly the same instant. So who gets to pick? Sometimes it's easy—the music stops and everyone claps along with the bass drum, or the lead singer starts one of those slow overhead claps to get the audience going—but often there is no such central sig-nal, and in those cases no one picks at all.

What happens is that when the crowd is close to synchrony, a few people, by random chance, may start clapping together. They're not doing it deliberately, and in isolation, their brief affair may last only a few beats. But that is long enough. Because they happen to be clumped together, they are temporarily louder than anyone else within earshot, and so they are more likely to drag someone else into synchrony with them than to be dragged apart. Hence, others are likely to join them, thus boosting their signal further and dragging in others still. Within seconds, they have become the nucleus around which the entire crowd has organized. But if an outside observer were to ask the ringleaders how they did it, in all likelihood they would be as surprised as anyone to discover their special status. Furthermore, if our observer were to rerun the experiment with exactly the same people in the same sta-dium, they would see the crowd coordinate around a different and equally arbitrary nucleus.

Much the same thing can be true of more complicated social

processes, like revolutions. In the end, the Serbian president cum dicta-
tor, Slobodan Milosevic, wasn't toppled by another political leader or
even by his army. Rather the driving force behind his fall was a loosely
organized and largely autonomous student movement called OTPOR,
which notably only acquired a coherent central leadership *after* it had
succeeded in mobilizing popular support. A traditional social network
analysis of the student movement would look at some of the principal
players in OTPOR and trace their involvement with each other, their
followers, and also outside organizations, and attempt to identify the
mechanisms by which they established themselves as central organizing
elements. But as we will see in chapter 8, when it comes to large-scale
coordinated social action, hindsight is not 20-20—in fact, it can be
actively misleading. Rather than leaders determining the events, quite
the reverse might have been true, with the particular sequence of events
and the peculiarities of their timing determining who it was that emerged
as leaders. In the simmering caldron of social discontent that was Serbia
in the summer of 2000, only a few, small, essentially random events were
required to bring the student movement and the population to a boil.
Many individuals were working to bring about the demise of Milosevic,
but only some of them later became leaders, and not necessarily because
they were *a priori* more special than the rest or even particularly well
positioned. Rather it was the playing out of the revolution itself that
determined where its center was, just like the nucleus in the clapping
crowd or the giant component of Erdős and Rényi's random graph.

So how does coherent global activity emerge from the interactions
of peers, without any centralized authority or control? As we will see in
the pages ahead, network structure is critical to this question, but so is
dynamics. As we have been using the term, however, *dynamics* really
has two meanings that are worth distinguishing, because each has
spawned an entire branch of the new science of networks. The first
meaning, and the one that will dominate the discussion throughout
chapters 3 and 4, is what we might call *dynamics of the network*. In this
sense of the word, *dynamics* refers to the evolving structure of the net-

work itself, the making and breaking of network ties. Over time, for example, we meet new friends and lose touch with old ones. Thus, our personal networks change, and the global structure of the social network to which we belong also changes. The static structures of traditional network analysis can be thought of as snapshots taken during this ongoing process of evolution. A dynamic view of networks, however, claims that existing structure can only be properly understood in terms of the nature of the processes that led to it.

The second meaning, which will occupy us in chapters 5 through 9, is what we might call *dynamics on the network*. From this perspective, we can imagine the network as a fixed substrate linking a population of individuals, similar to the traditional view of networks. But now the individuals are doing something—searching for information, spreading a rumor, or making decisions—the outcome of which is influenced by what their neighbors are doing and, therefore, the structure of the network. This is essentially the kind of dynamics that Steve Strogatz and I were thinking about when we veered off our project on crickets several years ago, and which for better or worse still dominates our own thinking about social processes.

In the real world, both kinds of dynamics are going on all the time. Social actors from revolutionaries to CEOs have to choose over and over again, not only how to respond to events as they perceive them but also with whom they will associate. If you don't like the way a friend is behaving, you can either try to alter his behavior or choose to spend your time with someone else. In response to a single scenario, the structure *of* the network could change, but so could the pattern of activity *on* the network. Furthermore, each kind of decision—each kind of dynamics—helps set the context in which subsequent decisions must be made. Your happiness affects your network, and your network affects your happiness. It's a complex dance, so in order to make some progress, we need first to understand each kind of dynamics on its own. Fortunately, in tackling these tasks, we have some giant shoulders on which to stand.

DEPARTING FROM RANDOMNESS

ANATOL RAPOPORT IS A MATHEMATICIAN BUT NOT ONE OF THE conventional variety. In a distinguished career spanning over half a century, he has made major contributions to psychology, game theory, and the evolution of cooperation, as well as epidemiology and the study of social networks. Back in the 1950s Rapoport was studying the spread of disease in human populations as part of a research group at the University of Chicago called the Committee on Mathematical Biophysics. In a time when most epidemiologists were focusing on disease models that ignored the social aspects of human interactions, the Chicago group understood that for some diseases the actual network is critical. In many circumstances, only by accounting for who interacts with whom can one determine how dangerous a particular outbreak of a disease might be.

We will return to this topic in later chapters because it is relevant not only to the spread of disease but also to the spread of information like rumors and computer viruses. What's important to mention now about Rapoport's early work is that although he came to the problem of network structure as a mathematician, he was heavily influenced by ideas from sociology, psychology, and biology. The reason, perhaps, is that he was relatively old—in his thirties—before he even started graduate school, on account of having served previously in the army and having fought in World War II. So by the time he got around to becoming a mathematician, he had already experienced a good many of the vicissitudes of life and possibly decided to incorporate them in his work.

Given an outbreak of a disease in a particular social network, Rapoport wanted to know how bad the situation could get. In other words, imagine that the disease is so incredibly contagious that virtually every person who comes into contact with an infected person becomes infected. How many people will be infected eventually? Well, it depends ultimately on how well connected the population is. If we are talking about the countryside of Central Africa, on the outskirts of the rain forest, where most people live in small, relatively isolated villages, we might imagine that an outbreak in a single village, although devastating to that village, will remain localized. If, however, we are talking about the North American continent consisting of huge dense populations connected in a multilayered web of air, road, and rail transportation, it is pretty clear that anything that virulent, starting anywhere, is going to explode. Is there, Rapoport asked, a critical level of connectivity between these two extremes at which the population transitions from a collection of small isolated populations into a single linked mass? This question should sound familiar—it is essentially the same question that Erdős and Rényi asked about communication networks and that led to the birth of random graph theory.

Indeed, Rapoport and his collaborators started off by examining randomly connected networks for much the same reasons as the Hungarian mathematicians, and although they used less rigorous methods, they reached rather similar conclusions (almost ten years before Erdős and Rényi!). Given his applied bent, however, Rapoport quickly saw through the analytical beauty of the random graph model and tried to address what he saw as its essential flaws. But if not random networks, then what? In the opening line of *Anna Karenina*, Tolstoy laments, "Happy families are all alike; every unhappy family is unhappy in its own way." In the same way, all random graphs are essentially the same, but nonrandomness is much harder to pin down. Do you, for instance, care that some friendships are asymmetrical or even unreciprocated? Should some relationships be recognized as more important than others? How does one account for the apparent preference of people to associate with others like themselves? Do most people have roughly the same number of friends, or do some people have many more friends than average? And how does one account for the existence of groups, inside which friendship ties, say, are dense but between which connections are relatively sparse?

Rapoport's group made some brave assaults on the problem, extending their work on random graphs to account for human characteristics like *homophily,* the "birds of a feather" tendency of like to associate with like, which characterizes not only college fraternities but also the personnel makeup of firms, clientele at stores and restaurants, and the ethnic character of neighborhoods. Homophily helps to explain why you know the people that you do—because you all have something in common—but one might also wonder how the people you know at present determine the people you will know in the future. Rapoport thought about this too, introducing the notion of *triadic closure.* In social networks, the basic unit of analysis is the *dyad,* a relationship between two people. But the next simplest level of analysis, and the basis of all group structure, is a triangle, or *triad,* which arises whenever an individual has two friends who are also friends of each other. Rapoport was not the first person to think about triads as the fundamental unit of group structure; the great German sociologist Georg Simmel had introduced the idea over half a century earlier. But what was revolutionary about Rapoport's work was that it included dynamics in the picture. Two strangers who possess a mutual friend will tend to become acquainted *in time;* that is, social networks (unlike random networks) will evolve in such a way that triads tend to close up on themselves.

In general, Rapoport conceived of the properties he was defining as *biases,* because each one took his models a step away from the assumption of pure randomness without dropping it altogether. Randomness is a powerful and elegant property that is often a perfectly adequate surrogate for the complicated, messy, and unpredictable things that happen in real life. Yet it clearly fails to capture some of the more powerful ordering principles that also govern the choices people make. So, Rapoport reasoned, why not balance these two sets of forces in a model? Decide which ordering principles you think are most important, and then imagine constructing networks that obey those properties but that are *otherwise* random. He called his new class of models *random-biased nets.*

The great power in this approach is that by treating networks as dynamically evolving systems, it avoided the central flaw in standard, static network analysis. Unfortunately in so doing, it also ran into two obstacles, one theoretical and one empirical, that turned out to be insurmountable. The first was data. Nowadays, in the aftermath of the Internet revolution, we are used to seeing data on and pictures of extremely large networks, including the Internet itself. Even more significantly, technology capable of recording social interactions electronically, from phone calls to instant messaging to on-line chat rooms, has increased the size of network data sets by several orders of magnitude in the last few years alone.

But data collection wasn't always like that. As recently as the middle 1990s, and certainly back in the 1950s, the only one way to get social network data was to go out and collect it by hand. That meant passing out surveys requesting that subjects recall their acquaintances and report the nature of their interactions with them. This method is not a very reliable way to obtain high-quality data, not only because people have a hard time remembering who they know without being suitably prodded, but also because two acquaintances may have quite different views of their relationship. So it can be hard to tell what is actually going on. The method also requires a lot of effort on behalf of the subjects and particularly the investigator. A much better approach is to record what it is that people actually do, who they interact with, and how they interact. But in the absence of electronic data collection, this technique is even more difficult to perform, in practice, than survey work. As a result, social network data, where it exists at all, tends to deal with small groups of people and is often restricted by the particular questions the researcher thought to ask beforehand. Basically then, Rapoport didn't have a target for his models, and if you don't even know what the world looks like, then it's very difficult to know if you have succeeded in capturing anything meaningful about it.

Rapoport, however, also faced an even more intractable problem. Although he understood the problem that he was *trying* to solve, he couldn't escape the fact that in the 1950s he really only had pencil and paper to work with. Even today, with our staggeringly fast computers, the analysis of random-biased nets is a hard problem. Back then it was virtually impossible. The fundamental difficulty is that as soon as you break the Erdős-Rényi assumption of every network connection arising independently of every other, it isn't clear anymore what depends on what. Triadic closure, for example, is only supposed to bias the network in a very particular way—by making cycles of length three (triads) more likely. That is, if A knows B and B knows C, then C is much more likely to know A than just anyone picked at random.

But once we start completing triads, we find that something else happens that we did not expect: we start to get cycles of other lengths as well. The simplest example of this unexpected dependency is demonstrated in Figure 2.3, in the first frame of which we consider four nodes, connected in a chain, that we assume to be part of a much larger network. Imagine now that node A gets to make a new link but has a strong inclination (bias) to connect to a friend of a friend. It is much more likely to connect to C than any other vertex, so let's assume that it does. Now we arrive at the second frame of the figure, where we can imagine that node D gets to choose a new friend. Again, node D has a bias to connect to a friend of a friend, and only two such nodes are available—nodes A and B—so D tosses a coin and chooses A, thus delivering us to the third frame. What has happened? All we specified was a preference to connect to a friend of a friend, or in other words to complete triads—cycles of length three—yet in so doing we have also created a cycle (ABCD) of length four.

There is nothing in our rule about cycles of length four—the bias only specifies triads—yet we will inevitably get them, along with cycles of other lengths that arise in the same incremental fashion. This occurs precisely because the construction of the network is a dynamic process, and the creation of each successive link takes as input the current state of the network, which includes all previously created links. The connection from D to A would probably not have happened had not the connection from *A* to *C* happened first. So not only can apparently very specific biases generate unintentional effects, but also the probability that any one event can happen at some point in the evolution of a network will depend in general on everything else that has happened up to that point.

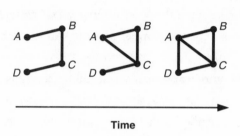

Time

Figure 2.3. Evolution of a random-biased net. A bias toward creating cycles of length three (triadic closure bias) also creates longer cycles. (Here *ABC* and *ACD* combine to make *ABCD*.)

Back in Rapoport's day, this realization was pretty much the end of the road, and reading his original papers you can see that he knew it. Perhaps if the University of Chicago group had had the kind of computers that we have today, they might have cracked the problem wide open, and network theory might have taken a very different route. But they didn't. Blinded by a lack of data and hobbled by a dearth of computational power, the theory of random-biased nets struggled as far as its few protagonists could take it with their mathematical intuition, and then effectively disappeared. It really was an idea for a future age, and like many such ideas, it had to do its time in purgatory.

HERE COME THE PHYSICISTS . . .

PHYSICISTS, IT TURNS OUT, ARE ALMOST PERFECTLY SUITED TO invading other people's disciplines, being not only extremely clever but also generally much less fussy than most about the problems they choose to study. Physicists tend to see themselves as the lords of the academic jungle, loftily regarding their own methods as above the ken of anybody else and jealously guarding their own terrain. But their alter egos are closer to scavengers, happy to borrow ideas and techniques from anywhere if they seem like they might be useful, and delighted to stomp all over someone else's problem. As irritating as this attitude can be to everybody else, the arrival of the physicists into a previously non-physics area of research often presages a period of great discovery and excitement. Mathematicians do the same thing occasionally, but no one descends with such fury and in so great a number as a pack of hungry physicists, adrenalized by the scent of a new problem.

In the decades since Erdős and Rapoport, while the sociologists were concentrating on static, structural explanations of networked systems, physicists were converging on a similar set of questions, albeit unintentionally and from the opposite direction. Instead of measuring structural properties of networks to understand the social roles of individuals and groups, the physicists effectively assumed perfect knowledge of individual-level properties, and by making very simple assumptions about structure, tried to work out the corresponding group-level properties. As was the case with sociology, the physicists' approach was driven by a desire to understand particular problems (although physical rather than social problems), a prime example of which was magnetism.

Most of us learned in high school science classes that magnets are composed of many smaller magnets, and that the field of the magnet being measured is actually the sum of all the fields of the smaller magnets. But each of these smaller magnets in turn is composed of smaller magnets still, and so on. Where does all this end? Where does the magnetic field ultimately come from? The answer, it turns out, comes from a deep equivalence between the electrical and magnetic fields, as first outlined by James Clerk Maxwell at the end of the nineteenth century. One result of Maxwell's unification of electromagnetism is that a spinning charged particle, like an electron, creates its own magnetic field, which, unlike an electrical field, possesses an inherent orientation that is determined by its direction of spin. A magnet, therefore, always has a north and south pole, whereas an electron, for example, has only a single negative charge. An important consequence of this fundamental physical fact is that a magnet can now be represented symbolically as a lattice of many tiny arrows, each of which corresponds to a spinning charged particle and is referred to as a *spin*. Magnetism can now be regarded as the state of the system in which all the spins (that is, the arrows) point in the same direction.

All other things being equal, magnetic spins prefer to be aligned with one another, so getting them all to point in the same direction might not sound problematic. But it can be, the reason being that the interactions between spins are sufficiently weak that each spin's orientation is affected only by the orientations of its immediate neighbors in the lattice. Global alignment, by contrast, requires that every spin somehow "know" about the direction of every other spin, even those that are far away. What tends to happen then is that groups of spins align locally, but neighboring groups turn out to point in opposite directions, and no one group has sufficient influence to flip the others. Even though the preferred state is one of global alignment, the system can get stuck in one these *frustrated* states, from which it can escape only through the application of an external magnetic field or by rattling it with additional energy. Magnetizing a piece of metal, therefore, typically requires placing it in the field of an existing strong magnet and then either heating it up or tapping it. However, if there is too much energy, all the spins will flip about at random regardless of what their neighbors or even the external field is trying to tell them. Therefore, in order to achieve global alignment, it is necessary to start the system off at a high temperature and then cool it down very slowly, typically in the presence of the external field.

One of the great triumphs of mathematical physics was figuring out exactly how the transition to magnetism works. Oddly enough, at the critical point of transition, all parts of the system act *as if* they can communicate with each other, despite their interactions being purely local. The distance over which individual spins can appear to communicate is usually called the *correlation length* of the system, and one way to think of the critical point is the state at which the correlation length spans the entire system. In this condition, known as *criticality*, tiny perturbations, which in any other state would be felt only locally, can propagate without bound throughout even an infinitely large system. The system therefore appears to exhibit a kind of global coordination but does so in the absence of any central authority. No center is needed when a system is at criticality because every site, not just some center, is capable of affecting every other site. In fact, because every site is, by definition, identical and they are all identically connected, there is no basis for one to be in charge of any other and therefore no basis for a center. As a result, no measure of centrality would be of any use in figuring out the root cause of the observed behavior. Rather, as was the case with our examples of random graphs and clapping crowds earlier, a series of small random events—events that would go unnoticed under normal conditions—can, at the critical point, push the system into a universally organized state, giving it the appearance of having been directed there strategically.

This all sounds a bit mysterious, but it represents the best understanding we have of how events at one scale can influence the systemic properties at another, even when every element of the system is only paying attention to its immediate neighbors. The excitement generated by this discovery has led the study of *spin systems* to become something of a cottage industry in physics, spawning many thousands of papers. Spin models are of deep interest to physicists, partly because they are so simple to state but mostly because they are relevant to so many phenomena—magnetic systems, the freezing of liquids, and other macroscopic changes of state such as the onset of superconductivity. And as you would have noticed if you've ever watched a cup of water freeze or hiked up through the snowline in the mountains, these changes of state are not steady and gradual, but sudden. One second it is raining, and the next, snowing. The magnet is either magnetized or it is not.

The transition through the critical point is in fact the physicists' version of a phase transition, one that turns out to be much like the transition between the disconnected and connected phases of a random graph. That we should be able to draw parallels between two such unrelated systems—the physics of a magnet and the connectivity of a mathematical object like a graph—should convey some sense of how deep the theory of phase transitions, and critical phenomena in general, really is. Regardless of whether we are discussing magnetization or the freezing of water into ice—procedures that involve completely different physics and even completely different materials—it turns out that *the nature of the corresponding phase transitions is the same!*

The observation that very different systems can exhibit fundamental similarities is generally referred to as *universality,* and its apparent validity presents one of the deepest and most powerful mysteries of modern physics. It is mysterious because there is no obvious reason why systems as different as superconductors, ferromagnets, freezing liquids, and underground oil reservoirs should have anything in common at all. And it is powerful precisely because they *do* have something in common, which tells us that at least some of the properties of extremely complicated systems can be understood without knowing anything about their detailed structure or governing rules. Classes of systems in which we can get away with ignoring many of the details are called *universality classes.* By knowing all the universality classes for a particular kind of model, physicists can make some very powerful statements about what can and cannot happen in different kinds of physical systems, again by knowing only the most basic facts about them. This is a tremendously hopeful message for anyone interested in understanding the emergent behavior of complex social and economic systems like friendship networks, firms, financial markets, and even societies.

One of the main obstacles that lies in the path of constructing simple models to describe such systems is that very little is understood about the fundamental rules that drive them. Einstein once said that physics deals with the easy problems. Not that physics is easy, but even with the hardest and most intractable problems like fluid turbulence and quantum gravity, physicists usually at least start with a reasonable idea of the governing equations. They may not be able to solve them or even to understand all the implications of the solutions they can find, but at least they can agree on what is to be solved in the first place. Economists and sociologists face a far bleaker outlook. In spite of two centuries of concerted effort, the rules governing individual social and economic behavior remain unresolved.

Perhaps the most successful attempt at a general theory of decision making to emerge from the social sciences is known as *rational expectations theory,* or simply *rationality.* Developed by economists and mathematicians in order to inject some scientific rigor into debates over human behavior, rationality has become the de facto benchmark against which all other explanations must be compared. Unfortunately, as we will see in later chapters, rationality makes a number of assumptions about human dispositions and cognitive capabilities that are so outrageous, several years of training in economic theory are required in order to take them seriously. More unfortunately still, no one appears to have come up with anything better.

In the 1950s, Herbert Simon and others proposed a far more reasonable-seeming version of rationality, called *bounded rationality,* that relaxes some of the more unlikely assumptions of the earlier theory without abandoning its more commonsense basis. Although even most economists agree that some version of bounded rationality must be true in reality, and although Simon won a Nobel Prize for his ideas, the problem is that once one starts to violate the assumption of perfectly rational behavior, there is no way to know when to stop. Just as there is no unique way to make random graphs nonrandom, there are so many ways in which rationality can be bounded that we can never be sure we have the right one.

The promise of universality is so enticing, therefore, precisely because it claims that we do not actually need to know the detailed rules of microlevel behavior and interaction—there are at least some questions that we may be able solve without them. This is a pretty big promise, so what's the problem? Universality has been understood for decades, and the theory of critical phenomena that has grown up around applications like magnetization and superconductivity is an extremely well-developed field of physics. Why then don't we understand epidemics, power failures, and stock market crashes?

The essential problem is that the physicists developed their tools to address physics questions; not social or economic ones, and sometimes that history gets in the way. Physicists, for example, are used to thinking about interactions between atoms in a crystal lattice. So when they try to apply their methods to human interactions, they tend to assume that people interact just like atoms do. The result is that the method looks very impressive and produces lots of elegant results, but it doesn't solve the actual problem *because it's not about the actual problem.* All the marvels of universality aside, some of the details do matter. And this is where people like sociologists come in. Because they have spent their lives studying the social world, they actually know a thing or two about how it works, and their insight is an indispensable element of any useful model.

As obvious as this last point may seem, it is endlessly surprising to most physicists, who rarely feel the need to consult anyone else before appropriating their problem. That will have to change if we want to make any real progress. Academics are a fractious bunch, rarely inclined to step across the boundaries of their disciplines for more than a polite hello. But in the world of networks, sociologists, economists, mathematicians, computer scientists, biologists, engineers, and physicists all have something to offer each other and much to learn. No one discipline, no single approach, has a stranglehold on a comprehensive science of networks, nor is that likely to happen. Rather, any deep understanding of the structure of real networks can come only through a genuine marriage of ideas and data that have lain dispersed across the intellectual spectrum, each a piece of the puzzle with its own fascinating history and insights, but none the key to the puzzle itself. As with jigsaw puzzles, the key is the way in which all the parts interlock to form a single unified picture. That picture, as we will see in the chapters ahead, is far from complete, but due to the efforts of many researchers across multiple fields, and with several distinguished lineages of intellectual endeavor to build on, it is finally starting to come into focus.

CHAPTER THREE

Small Worlds

●—●—●—●—●—●

Back when steve strogatz and i started our work together,
we didn't know any of this. Neither of us had the foggiest idea about
Rapoport or Granovetter, or really anything about social networks at
all. We both knew some physics—in fact, I had majored in it at college.
But college had been a military academy, and what little actual knowledge had seeped through the cracks of my officer training, outdoor
adventures, and generally earthly preoccupations of a young man's life
in the Navy seemed remote and of little immediate relevance. Graph
theory was also a mystery. Effectively a branch of pure mathematics,
graph theory can be divided roughly into two components—the almost
obvious and the utterly impenetrable. I learned the obvious stuff out of
a textbook, and after some futile struggling with the rest, convinced
myself that it wasn't very interesting anyway.

All this profundity of ignorance left us in something of an awkward
place. We were reasonably certain that someone must have thought
about this problem before, and we worried that we might waste a lot of
time reinventing the wheel. But we also thought that if we went out
looking for it, we might get discouraged by how much had already been
done, or else trapped into thinking about the problem from the same
perspective and so get stuck on the very same things that other people

had. After I spent a month at home in Australia to think about it, we met in Steve's office in January 1996 and made up our minds: we would go it alone. Telling almost nobody and reading virtually nothing, we would drop the crickets project and have a go at building some very simple models of social networks to look for features like the small-world phenomenon. No doubt feeling that he needed to protect me from myself, Steve insisted that we give it only four months—a single semester—after which, if we hadn't made some significant progress, we would concede defeat and return to the crickets. At worst, my graduation would be delayed by a semester, and if it would make me happy, well, why not?

WITH A LITTLE HELP FROM MY FRIENDS

AT THAT POINT I HAD BEEN LIVING IN ITHACA FOR JUST OVER TWO years and was beginning to feel like I had a new home with new friends, but still felt closely connected to my old ones. It occurred to me, however, that if you asked the average Cornell student how close he or she felt to a random person in Australia, the answer would probably be "not very." After all, most of my friends in America had never met another Australian before, and few of my Australian friends knew any Americans. The two countries are on virtually opposite sides of the planet, and despite a certain cultural similarity and a good deal of mutual fascination, are viewed by most of their inhabitants as being almost impossibly distant, even exotic. Nevertheless, at least some small group of Americans and some small group of Australians actually were very close to each other, although they might not have known it, by virtue of a single common friend—me.

A similar state of affairs applied on a smaller scale between my different groups of friends at Cornell. I belonged to the Department of Theoretical and Applied Mechanics, which was a small graduate department in which there were more foreign students than Americans. I spent an awful lot of time in that department and got to know the other graduate students pretty well. But I also taught rock climbing and skiing in Cornell's outdoor education program, and most of the friends from Cornell that I still have today were fellow outdoor instructors or students. Finally, I had lived in a big dormitory in my first year and had made some good friends there. My classmates knew each other, my dorm friends knew each other, and my outdoor friends knew each other. But the different groups were all quite, well, *different*. Without me to come and visit, my climbing friends, for example, would have precious little reason ever to venture into my department in Kimball Hall, and (with some justification) tended to regard engineering graduate students as something of a breed apart.

That two people can share a mutual friend whom each regards as "close," but still perceive each other as being "far away" is a facet of social life at once commonplace and also quite mysterious. As we will see in chapter 5, this paradox lies at the heart of the small-world problem, and through its resolution we can understand not only Milgram's results but also a number of other network problems that on the surface have nothing to do with sociology at all. That will take some more work, however. For now, it's enough to say that we don't just have friends, rather we have groups of friends, each of which is defined by the particular set of circumstances—some *context*, like our college dorm or our current workplace—that led to our getting acquainted. Within each group there will tend to be a high density of interpersonal ties, but ties between different groups will typically be sparse.

The groups, however, *are* connected by virtue of individuals who belong to more than one group. In time, these *overlaps* between groups may grow stronger, and the boundaries between them blur, as people from one group start to interact with people from another via the intermediation of a mutual friend. Over the years that I spent at Cornell, my different groups of friends eventually met each other and occasionally became friends themselves. Even some of my Australian friends came to visit, and although they didn't stay long enough to form any lasting relationships, the boundary between the two countries is now, in some small way, less distinct than it was.

After thinking through these ideas a number of times, and a good deal of wandering around a very cold Cornell campus, Steve and I decided that there were four elements we wanted to capture in our model. The first was that social networks consist of many small overlapping groups that are densely internally connected and that overlap by virtue of individuals having multiple affiliations. The second was that social networks are not static objects. New relationships are continually being forged, and old ones abandoned. Third, not all potential relationships are equally likely. Who I know tomorrow depends at least to some extent on who I know today. But the final feature was that we occasionally do things that *are* derived entirely from our intrinsic preferences and characteristics, and these actions may lead us to meet new people who have no connection to our previous friends at all. My decision to move to America was driven solely by my desire to go to graduate school, and I didn't know a soul when I got there, nor did anyone else that I knew. Likewise, my decision to teach climbing was unaffected by my choice of department, as was the dorm I lived in.

In other words, we do what we do in part because of the position we occupy in our surrounding social structure and in part because of our innate preferences and characteristics. In sociology, these two forces are called *structure* and *agency,* and the evolution of a social network is driven by a trade-off between the two. Because agency is the part of an individual's decision-making process that is *not* constrained by his or her structural position, actions derived from agency appear as random events to the rest of the world. Of course, decisions like moving to another country or going to graduate school are derived from a complicated mixture of personal history and psychology and so are not random at all. But the point is that as long as they aren't determined explicitly by the current social network, we can treat them *as if* they are random.

Once these apparently random affiliations have been made, however, structure reenters the picture, and the newly created overlaps become the bridges over which other individuals can cross and form additional affiliations of their own. The dynamic evolution of relationships in a social network, therefore, is driven by a balance of conflicting forces. On the one hand, individuals make what seem like random decisions to launch themselves into new social orbits. And on the other, they are constrained and enabled by their current friendships to reinforce the group structures that already exist. The $64,000 question was, How important is one relative to the other?

Obviously we didn't know. And furthermore, we were pretty sure that nobody else would either. The world, after all, is a complicated place, and it is precisely this kind of uncertain, hard-to-measure balance between conflicting forces that makes it so. Fortunately, empirical tangles like this are precisely the places where theory comes into its own. Instead of trying to establish the balance between individuals willfulness and social structure—between randomness and order—that exists in the *actual* world, we could ask the question, What can we learn by looking at *all possible* worlds? In other words, think of the relative importance of order and randomness as a parameter that we can tune in order to move through a space of possibilities, in much the same way that the tuning knob on an old-fashioned radio allows us to scan the spectrum of radio frequencies.

At one end of the spectrum, individuals *always* make new friends through their current friends, and at the other end, they *never* do. Neither extreme is very realistic, but that was the point—by picking unreasonable extremes we hoped that somewhere in the messy middle ground we would find a believable version of reality. And even if we couldn't specify exactly where that point might be, our hope was that much of what lay in between the two extremes would be, in some well-defined sense, the same. What we were looking for was not a single kind of network to hold up as a model of a social network, but in the spirit of universality, a *class* of networks, each of which might differ in detail from all the others but whose essential properties didn't depend on those details.

Casting around for the right kind of model took some time. The notion of group structure that we had started off with turned out to be harder to capture in a precise manner than we had expected. But eventually the breakthrough came. As always, I immediately rushed down the corridor to Steve's office and hammered on the door until he gave up trying to do whatever it was he was trying to do, and let me in.

FROM CAVEMEN TO SOLARIANS

Perhaps it's not surprising, but as a boy I was a keen Isaac Asimov fan. In particular I read and reread his two most famous series: the *Foundation* trilogy and the *Robot* series. Oddly enough, the psychohistory of Hari Seldon, the chief protagonist in *Foundation*, was probably my earliest exposure to the idea of emergence in social systems. As Seldon puts it, although the behavior of individuals is hopelessly complex and unpredictable, the behavior of mobs, and even civilizations, is amenable to analysis and prediction. As fantastical as it was when conceived in the early 1950s, Asimov's vision is remarkable in its anticipation of much that the study of complex systems attempts to do today. It was, however, the *Robot* series that I wanted to talk to Steve about.

In *The Caves of Steel*, the first book of the series, detective Elijah Baley sifts through a murder mystery on a futuristic Earth built entirely underground. While doing so, he also contemplates the mysteries of his own life and his relations with his fellow man. Amid the teeming masses of humanity packed into their caves of steel, Baley knows a small, tight-knit group of people very well and almost nobody else. Strangers do not talk to one another, and interactions between friends are physical and personal. In the sequel, *The Naked Sun*, Baley is sent on assignment to the colonized planet of Solaria, which much to his discomfort lies at the opposite end of the social interaction spectrum. Unlike native Earthmen, Solarians live on the surface of a sparsely populated planet. They inhabit enormous estates in solitary isolation, accompanied only by robots, and interact with each other (even their spouses) virtually, via what amounts to a global teleconferencing facility. Back on Earth, life is lived in the security of interlocking and mutually reinforcing ties, and initiating a relationship with a random stranger would be inconceivable. But on Solaria, all interactions are equally accessible, and prior relationships are relatively unimportant to the establishment of new ones.

Imagine then, two worlds—a world of caves and a world of random, independent relationships—and ask, how do new relationships get formed in each one? Specifically, think about the likelihood of meeting a particular, randomly selected person as a function of how many mutual friends you and that person currently possess. In a caveman world, an absence of mutual acquaintances suggests you live in different "caves," so you will probably never meet. But if you have even one common friend, the implication is that you live in the same community, move in the same social circles, and so are extremely likely to become acquainted. Obviously, this would be a strange place in which to live, but again the point is to find the extremes. At the other extreme, akin to Solaria, your social history is irrelevant to your future. Even if two people happen to have many friends in common, they are no more or less likely to meet than if they have none.

Each of these general principles for choosing new friends can be expressed more precisely in terms of what we might call *interaction rules.* In the universe of our model, we could construct a network of nodes connected by social ties (let us imagine them to be friendships, although they need not be), and then let the network evolve in time as individuals make new friends according to a specified interaction rule. The two extreme types of worlds, the caveman world and Solaria, for example, can be captured by the rules in Figure 3.1. We can see that the tendency of two people to become friends is determined by the number of friends they currently have in common, but that the precise way in which it is determined varies dramatically from one rule to the other. The top curve corresponds to a caveman world because as soon as two individuals have even a single mutual friend, they immediately display a strong tendency to become friends themselves. The bottom curve, by contrast, corresponds to the Solaria world in which even a very large

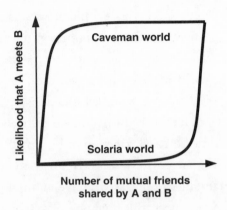

Figure 3.1. Two extreme kinds
of interaction rules. In the top
curve (caveman world), even a
single mutual friend implies that
A and B are highly likely to meet.
In the bottom curve (Solaria
world), all interactions are
equally unlikely, regardless of
how many friends A and B share.

number of mutual friends has precious little effect on people's tendency to interact. So under almost all circumstances, they interact randomly.

The great advantage of formalizing the rules of network evolution in this way is that, as we can see in Figure 3.2, a continuum of *intermediate* rules can be defined as curves that lie between the two extremes. Each one of these rules expresses the tendency of two individuals to become friends as a function of how many mutual friends they have at that moment, but they vary in the extent to which mutual friends matter. Mathematically, this entire *family* of rules can be expressed in terms of an equation that contains a single *tunable parameter*. By adjusting, or *tuning*, the parameter betwen zero and infinity, we can pick out one of the interaction rules in Figure 3.2, and then build a network that evolves according to this rule. What we have created is a mathematical model of a social network. Because this was the first network model Steve and I ever built, we called it, for want of a better name, the *alpha model,* and so the parameter driving its behavior became *alpha.*

Although we didn't know it at the time, the alpha model was very close in spirit to Anatol Rapoport's random-biased nets. And like Rapoport, we quickly found it impossible to solve anything using only pencil and paper. Fortunately for us, five decades of technological development had finally yielded computers fast enough to do the job by brute force. Indeed, in many ways network dynamics problems are ideal grist for the mill of computer simulation. Very simple rules, at the level of individual actions, can generate bewildering complexity when many such individuals interact over time, each making decisions that necessarily depend on the decisions of the past. Often the results are highly counterintuitive, and pencil-and-paper math very rarely works on its own. Computers, however, love this stuff, the endless, mindless,

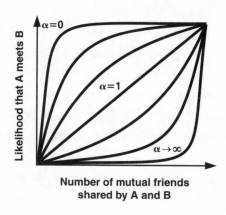

Figure 3.2. Between the two extremes, a whole family of interaction rules exists, each one specified by a particular value of the tuneable parameter alpha (α). When $\alpha = 0$, we have a caveman world; when α becomes infinite, we have Solaria.

blindingly fast iteration of simple rules being precisely what every computer was born to do. In the same way that physicists do experiments in the lab, computers have enabled mathematicians to become experimentalists, testing their theories in a multitude of imaginary laboratories where the rules of reality can be adjusted at will.

But what sorts of things were we supposed to test? Remember that the problem we wanted to understand—the origin of the small-world phenomenon—seemed to depend on the presence of two apparently contradictory properties of social networks. On the one hand, the network should display a large *clustering coefficient*, meaning that on average a person's friends are far more likely to know each other than two people chosen at random. On the other hand, it should be possible to connect two people chosen at random via a chain of only a few intermediaries. Hence, even globally separated individuals will be joined by short chains, or *paths* in the network. Each of these properties is trivial to satisfy on its own, but it wasn't at all clear how they could be combined. The caveman world of Elijah Baley, for example, is clearly highly clustered, but our intuition would suggest that if all the people we know tend to know only each other, it would be very hard to connect ourselves through them to the rest of the world in only a few steps. All that local redundancy might be good for group cohesion, but it's clearly no help in fostering global connectivity. By contrast, the Solaria world is much more likely to display short network path lengths. In fact, when people interact purely at random, it is a standard result from graph theory that on average the typical length of a path between any two of them will be short. However, it is also easy to show that in a random graph, the probability that any of our friends will know each other becomes negligible in the limit of a very large global population; hence, the clustering coefficient will be small. Our intuition, therefore, suggests that either the world can be small or it can be clustered, but it can't be both. Computers, however, are unconcerned with intuition.

SMALL WORLDS

USING PATH LENGTH AND CLUSTERING AS OUR PROBES, WE STARTED to build our *alpha networks* in the computer, first constructing them and then implementing some standard algorithms to measure the corresponding statistics. The programming required was mostly elementary, but I had to teach myself the language in the process, so the resulting code was ugly and sluggish, and I frequently spent many hours trying to track down some bug that killed my program after running happily for a day or more. Computer simulation may be less messy than the real world, but it can still be a pain. After a frustrating month or so, however, we at last had some results to ponder.

At first it seemed that our intuition had been correct. When the value for alpha was low, meaning that the nodes had a strong preference to connect only to friends of friends, the resulting graphs tended to be

highly clustered. So much so, in fact, they actually fragmented themselves into many tiny components, or *caves*. Within each cave, everyone was well connected to one another, but between different caves no
connections existed at all. This result was actually an inconvenience,
because when networks are fragmented in this fashion, it is difficult to
define the distance between nodes in different fragments. Fortunately
it *is* possible to define a sensible notion of path length that can account
for network disintegration. In the simplest such modification, one
measures the shortest path length between pairs of nodes exactly as
before, but only computes the average over pairs that lie in the same
connected component. The result, as shown in Figure 3.3, is that the
typical path length is small when the alpha value is low and it is small
when the alpha value is high, but the path length spikes upward when
the alpha value is somewhere in the middle. The explanation is that for
low alpha, the graph is highly fragmented, but because the average is

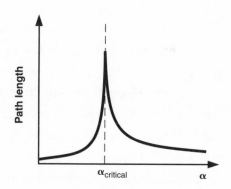

Figure 3.3. Path length as a
function of alpha (α). At the
critical alpha value, many small
clusters join to connect the
entire network, whose length
then shrinks rapidly.

computed only over nodes in the same connected components (the
caves), the small size of the components themselves yields short path
lengths. This is the *Caves of Steel* world—the people who can be reached
can be reached easily, and the people who can't be reached easily can't be
reached at all. When the alpha value is high, by contrast, the graph is
more or less random. As a result, it is now connected into a single universal component, and the typical separation between any pair of nodes is

small, just as we know is true for random graphs. This is the Solaria world, in which *everyone* can be reached with roughly the same ease.

The spike in the middle of Figure 3.3 is where all the interesting behavior is. To the left of the spike, as alpha increases, the fragments are rapidly joining together, resulting in a large increase in the apparent path length. The world is getting bigger, but it is doing so because the previously isolated components are starting to get connected. It is harder to reach people, on average, but more and more people can be reached. To the right of the spike, all the components in the network have been connected into a single entity, and now the average path length starts to collapse rapidly as the interaction rule becomes more random still. Right at the spike is a critical point, a phase transition very similar to the one discussed for random graphs, at which everyone becomes connected but the typical path length between pairs of individuals tends to be very large. At the peak of the spike, in a network of, say, a million people, each of whom has a hundred friends, the typical path length would be in the thousands. A network in which you are a few thousand handshakes from the president is clearly the antithesis of a small world. But—and this is important—such a world is inherently unstable. Almost as soon as the phase transition has occurred, and the network has become globally connected, the average path length starts to drop like a stone, rapidly approaching its eventual minimum. Although it seemed mysterious at the time, it was this surprisingly rapid drop in the length that turned out to be critical.

The clustering coefficient also displayed some unexpected behavior, first rising to a maximum for some low value of alpha, and then also falling rapidly, just like the average path length. What was more interesting, however, was the location of this transition relative to the corresponding transition in path length. Because we expected, on the one hand, highly clustered graphs with large characteristic lengths and, on the other hand, poorly clustered graphs with small characteristic lengths, we expected that the transition of both statistics might also correspond to each other. Instead, as Figure 3.4 shows, the length started to plummet just as the clustering was reaching its maximum value.

At first we assumed there was an error in the code, but after some careful checking and some vigorous head scratching, we realized that what we were staring at was the very small-world phenomenon we were looking for. In the universe defined by our model, here was a regime in

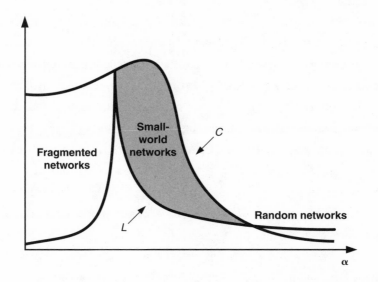

Figure 3.4. Comparison between path length (*L*) and clustering coefficient (*C*). The region between the curves, where *L* is small and *C* is large (shaded), represents the presence of small-world networks.

which networks displayed the high local clustering of disconnected caves but were connected such that any node could be reached from any other in an average of only a few steps. We called this class of networks *small-world networks*, which wasn't perhaps the most scientific of labels but had the great advantage of being catchy. Small-world networks have since received a lot of attention and while somewhere in all the fuss the original alpha model has been largely forgotten, it still has a few things to teach us about the world.

The first thing the alpha model tells us is that either the world will be fragmented into many tiny clusters, like isolated caves, or it will be connected in a single giant component within which virtually anyone can be connected to anyone else. It is not possible, for example, to have two, or even a few, large components between which the world is evenly divided. This result may seem surprising because the world often appears divided along geographical, ideological, or cultural lines into a small number of large and incompatible factions—East and West, black and white, rich and poor, Jewish, Christian, and Muslim. Although these cleavages although such dichotomies may drive our perceptions, and thereby affect our actions in important ways, what the alpha model tells us is that they do not apply to the network itself. We are either all connected or not at all connected—there really isn't anywhere in-between.

Furthermore, it turns out that the highly connected state is overwhelmingly more likely than the highly fragmented one. Our parameter alpha, remember, represents a balance between the constraints of social structure and the freedom of individual agency. At the moment, alpha is a hard parameter to interpret, so it is unclear exactly what a particular value means in terms of the real world. Once we learn a little more about networks, however, it will become clear that just a little bit of agency goes a long way. The result is that the world we live in almost certainly lies to the right of the peak in Figure 3.4, implying that each of us can be connected to everyone else. In fact, the model makes an even stronger statement than this. Because the drop-off on the right side of the peak is so rapid, not only is it likely that the world is globally connected, but it is almost certainly the case that the world is *small* in the sense that almost any pair of individuals can be connected via a short chain of intermediaries. This result may also come as a surprise to many of us who spend most of our lives interacting with a relatively small group of people—friends, family, and coworkers—who are often quite similar to ourselves. Even educated, privileged people can feel isolated within their little communities. They may not be unhappy about it, but they are still likely to feel impossibly distant from the large majority of the world that is completely unlike the relatively tiny portion they actually know. So how is it that despite this (quite real) perception, we really *are* all connected?

The resolution of the paradox is that the clustering coefficient does not drop off nearly as rapidly as the path length. No matter what the network is like on a global scale—whether fragmented or connected, large or small—the clustering coefficient is almost certainly going to be high. Individuals, therefore, have severe limitations imposed on what they can deduce about the world based on what they can observe. A well-known aphorism contends that all politics is local, but really we should say that all *experience* is local—we only know what we know, and the rest of the world, by definition, lies beyond our radar screen. In social networks, the only information we have access to, and therefore the only data we can use to make assessments of the world, lies in our *local neighborhood*—our friends and acquaintances. If most of our friends know each other—if our local neighborhood is highly clustered—and if everyone else's neighborhood is likewise clustered, then we tend to assume that not all these clusters can be connected.

But they can be, and that's why the small-world phenomenon is so counterintuitive—it is a global phenomenon, yet individuals are capable only of local measurements. You only know who you know, and maybe *most* of the time, your friends know the same sort of people you do. But if just *one* of your friends is friends with just *one* other person who is friends with someone not like you at all, then a connecting path exists. You may not be able to use that path, you may not know it's there, and finding it may be difficult. But it *is* there. And when it comes to the propagation of ideas, influence, or even disease, that path can matter whether you know about it or not. Just like Hollywood, who you know is important, but now there's more to the story: who your friends know, and who *those* people know can be important as well.

AS SIMPLE AS POSSIBLE

THE ALPHA MODEL WAS AN ATTEMPT TO UNDERSTAND HOW SMALL-world networks can be generated in terms of the rules that people follow when making new friends. But once we knew the small-world phenomenon was possible, we wanted to figure out exactly what generated it. It didn't seem adequate simply to conclude that the effect we were seeing was a function of our parameter alpha, because we didn't really know what alpha was, and therefore what any particular value of alpha might signify. As simple as the alpha model was, it was still too complicated, so we decided that if we really wanted to understand what was going on, we needed to follow Einstein's famous dictum, making it "as simple as possible, and no simpler." So what was the simplest model that could replicate the small-world phenomenon? And what, in its simplicity, would it tell us that the alpha model hadn't? What we set out to do with our second—hence, *beta*—model was to abandon even the most superficial pretense of modeling social networks and treat structure and randomness in as abstract a manner as possible.

In physics, as we already discussed, the interactions between the elements in a system often take place in a lattice. Lattices are particularly convenient objects to study because every site in a lattice is identical to every other site, so once you know your own location, you know where everyone else is also. This is why grid systems are so popular for laying out roads in cities or cubicles in a large office space: they are extremely easy to navigate. The only slightly tricky cases are those that lie on the boundaries, because those sites have fewer interactions than their interior counterparts. This asymmetry is easily fixed (mathematically, if not in the office) by "wrapping" the space around on itself such that opposite sides join up. Thus, a straight line becomes a ring, and a square lattice becomes a torus (Figure 3.5). Rings and tori are called *periodic lattices* because there is no longer any boundary across which one can exit the space. Any point moving from site to site on the lattice is destined to keep going around and around, *periodically*, just like enemy ships on the old Space Invaders games.

Periodic lattices, therefore, seemed like an entirely natural class of networks to embody the notion of *ordered* interactions. At the other extreme, a random network appeared to be the embodiment of *disordered* interactions. And although they are not nearly as simple as lattices, random networks are quite well understood also. More specifically, whereas the properties of a periodic lattice can be specified exactly, the properties of a random graph can be specified statistically. Imagine two trees of the same species and roughly the same size growing next to each other in the same soil. Obviously they will never be exactly the same, yet it is equally clear that they are, in some sense, interchangeable. Random graphs are predictable in roughly the same way: given any pair of sufficiently large random graphs with the same parameters, no *statistical* test would be able to tell them apart.

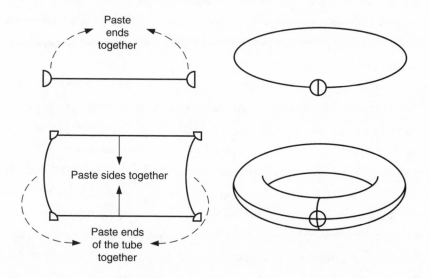

Figure 3.5. A lattice can be made periodic by joining up its opposite edges. In the top diagrams, a one-dimensional lattice (left) becomes a ring (right). In the bottom diagrams, a two-dimensional lattice (left) becomes a torus (right).

Thus, a network could be considered *ordered* to the extent that it resembled a lattice and *disordered* to the extent that it resembled a random graph. All we needed to do was find a way to tune each network between complete order and complete disorder in a way that it traced through all the various intermediate stages. Although these partly ordered, partly random networks are still hard to understand in purely mathematical terms, they are cannon fodder for the computer, and we quickly developed a simple algorithm for building them. Picture a regular lattice, like the one shown on the left of Figure 3.6, in which each node is connected to a fixed number of nearest neighbors on the ring. In this arrangement, for example, if you have ten friends, you know the five immediately on your left and the five on your right. As with the extremes of the alpha model, this kind of social network is pretty strange—it would be like everyone standing in a ring holding hands, and the only way to communicate is by shouting to someone within earshot. But remember that we are not trying to build social networks here—only to interpolate between ordered and disordered networks in some simple fashion.

Now imagine that we can introduce cellular phones. Instead of talking to one of your neighbors, you now have a phone that connects you directly to someone else, chosen at random from the entire network. In Figure 3.6, this equates to choosing a link at random and *rewiring* it. That is, delete a link between A and B, and then keeping A's end fixed, choose a new friend, B_{new}, at random from the ring. In practice, what we do is pick a value of beta (our new tunable parameter) between zero and one, and then systematically visit every link in the lattice and randomly rewire it with probability beta. So if beta is zero, then no rewiring takes place (no one has a cell phone), and we end up precisely where we started—a perfectly regular lattice. At the other extreme, when beta is one, every single link gets rewired, and the result is a highly disordered network (right diagram in Figure 3.6) that resembles a random graph.

These two extremes of the beta model are much simpler to understand than the corresponding extremes of the alpha model, which, remember, were defined in terms of the interaction rules governing individual nodes. Dynamically grown networks, like the alpha model, are generally hard to analyze because it is frequently unclear exactly what it is about the underlying behavioral rules that generates the observed structure. Perhaps even more important, many kinds of underlying behavioral rules might conceivably generate the same kinds of structural features in the final network, and this was the problem that most interested us. We knew how to generate small-world networks dynamically. Now we wondered to what extent they could exist independently of how they were created.

Besides being at opposite ends of the order-randomness spectrum, how else do lattices differ from random graphs? First of all, a ring lattice is "large" in the sense that when it consists of very many people, the typical number of steps—the path—between any two tends to be large. Imagine, for instance, that you want to pass a message to someone on the opposite side of the ring in the left diagram in Figure 3.6. Say that the ring consists of a million people and each person has one hundred friends, fifty on their left and fifty on their right. The fastest way to transmit your message is to shout it out to the fiftieth person on your left, and tell him to pass it on. At which point, he can shout to *his* fiftieth friend on the left, and tell her the same thing. In this way, your message skips around the ring in hops of fifty people at a time, taking fully ten thousand such steps to get all the way to its destination. Not every-

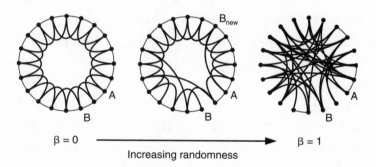

$\beta = 0$ $\longrightarrow$ $\beta = 1$

Increasing randomness

Figure 3.6. Construction of the beta model. The links in a one-dimensional, periodic lattice are randomly rewired with probability beta (β). When beta is zero (left), the lattice remains unchanged, and when beta is one (right), all links are rewired, generating a random network. In the middle, networks are partly ordered and partly random (for example, the original link from A to B has been rewired to B_{new}).

one is as far away from you as the person on the opposite side, but the average distance is still about five thousand degrees of separation, a far cry from six. A ring lattice is also highly clustered for the simple reason that the person next to you, by virtue of the lattice structure, knows almost all the same people you do. Even the person at the extreme edge of your friendship circle still knows half of your friends, so the clustering coefficient, averaged over all your friends, is roughly midway between one-half and one, or three-fourths.

By contrast, a fully randomly rewired graph exhibits negligible clustering. In a very large network, the chance of you randomly rewiring to two people who subsequently are randomly rewired to each other is vanishingly small. For this same reason, a random graph will automatically be *small* in the same way that a lattice is large. Remember our very first thought experiment concerning the small-world phenomenon? If I know one hundred people, and each of them knows one hundred people, then within two degrees of separation I can reach ten thousand people, within three degrees I can reach almost one million, and so on. The absence of clustering means no wasted or redundant connections—every new additional connection reaches out into new territory—so the rate of growth of my acquaintance network is as rapid as possible. As a result, I can reach everyone else in the network in only a few steps, even when the population is very large.

So what happens in the middle? When the rewiring probability is small, as in the center of Figure 3.6, the resulting object looks very much like a regular lattice, but it has a few random, long-range connections. What difference do they make? If you look at the clustering coefficient, a few random links makes very little difference at all. For each random rewiring, you know one fewer of your neighbors and you have instead one additional friend who doesn't know anyone else you know. Nonetheless, most of your friends still know each other, so the clustering remains high. The path length, however, changes dramatically. Because the links get rewired uniformly at random, and because in a big lattice there are many more lattice sites far away from you than nearby, the chances are that you will get connected to someone who used to be a long way away. Hence, random links tend to create *short-cuts,* and shortcuts, as the name suggests, serve the role of contracting path lengths between previously distant nodes.

Returning to the cell phone analogy, instead of having to pass a message to the opposite side of the ring in hops of fifty, now both you and your intended recipient have phones, shrinking the distance between the two of you, in one fell swoop, from several thousand to one. And not only that. If you want to pass a message to your new friend's friends, you can reach them in only two steps. Furthermore, their friends can talk to your friends, and friends of their friends can talk to friends of your friends, in only a few short hops, all via you and your connection on the other side of the world. Roughly speaking, this is how the small-world phenomenon works. In a large network, every random link is likely to connect individuals who were previously widely separated. And in so doing, not only are they brought together, but also large chunks of the rest of the network are made much closer.

The key observation is that only a few random links can generate a very large effect. As you can see in Figure 3.7, when beta increases from zero, the path length drops off a cliff, plummeting almost too fast to distinguish it from the vertical axis. At the same time, in shrinking the distance between many pairs of nodes, each shortcut reduces the marginal effect of any subsequent shortcuts. So the rapid plunge in length decelerates almost as soon as it has begun, converging gently to its random graph limit. For this simple model, one surprising result is that on average, the first five random rewirings reduce the average path length of the network by one-half, *regardless of the size of the network.* The bigger the network, the greater the effect of each individual random link so the impact of adding links becomes effectively independent of size. The law of diminishing returns, however, is just as striking. A further 50 percent reduction (so that now the average path length is at one-fourth of its original value) requires roughly another fifty links—roughly ten times as many as for the first reduction and for only half as much overall impact. Subsequent reductions require many more random links—a great deal more disorder—for ever smaller effects. Meanwhile, the clustering coefficient, like the tortoise racing the hare, continues its slow and steady descent, eventually catching up with the characteristic length in the limit of complete disorder.

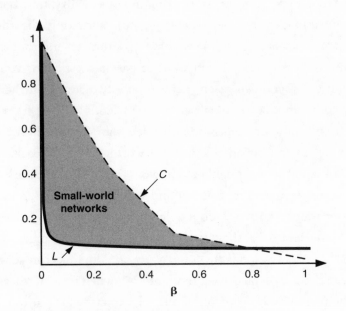

Figure 3.7. Path length and clustering coefficient in the beta model. As with the alpha model (see Figure 3.4), small-world networks exist when path length is small and the clustering coefficient is large (shaded region).

The net result is that we once again find a wide interval in the space of networks between complete order and complete disorder, in which local clustering is high and global path lengths are small. These are our small-world networks. As with the alpha model, individuals located somewhere in a small-world network *cannot tell* what kind of world they're living in—they just see themselves as living in a tight cluster of friends who know one another. The consequences of this statement are important, as we will see in later chapters when we learn about the spread of disease and computer viruses, on the one hand, and the search for information in large organizations and peer-to-peer networks, on the other.

But the beta model also tells us something deeper, because it helps us resolve the problem of the mysterious alpha parameter from our first model. The problem with alpha, remember, was that it was impossible to interpret in terms of the network itself. When alpha is small (the caveman world), we build networks in which people with even a single mutual friend have a strong tendency to become friends themselves. And when alpha is very large (the Solaria world), people tend to meet randomly whether they have a mutual friend or not. But as we saw, it's generally impossible to predict exactly what kind of network will result from a given value of alpha, especially the values in the middle region that generate the most interesting behavior.

Now we can understand it. Alpha determines the probability that the finished network will exhibit long-range, random shortcuts, and it is the shortcuts that do all the work. The beauty of this result is that we can now generate the shortcuts in almost any way we want to—by simulating the social process of networking, as in the alpha model, or simply by creating them with some probability, as in the beta model—and we will get more or less the same result. And much the same is true of the clustering. We can simply put it there, as we did with the lattice in the beta model, or else we can let clustering arise naturally through the iteration of a rule for making new friends via current friends. Either way, as long as we have a way of generating clustering and a way of allowing shortcuts, we will always get a small-world network.

So even though the beta model was kind of silly in that no real system would actually look like it, the message it delivered wasn't silly at all. What it told us was that small-world networks arise from a very simple compromise between very basic forces—order and disorder—and not from the specific mechanisms by which that compromise is brokered. At that point it dawned on us that small-world networks should show up not just in the social world, from which the idea originated, but in all kinds of networked systems.

THE REAL WORLD

As OBVIOUS AS IT SEEMS NOW, THE REALIZATION THAT SMALL-world networks should arise in all sorts of networked systems was actually quite a breakthrough for us, as up until that point we had really been thinking about the problem purely in terms of social networks. On a more practical level, it also opened up the possibility of finding some data with which to check our prediction. Remember that one of the big problems in studying the small-world phenomenon, and the reason we ended up adopting the tuning-between-order-and-random-ness approach, was that empirical verification of the phenomenon itself seemed completely unfeasible. Who could get that kind of network data? Now, however, our range of acceptable network data was expanded dramatically. Essentially any large network would do, as long as it was sufficiently well documented. In practice, this last condition meant that it had to be available electronically, a requirement that nowadays seems trivial. Back in the Internet dark ages of 1997, however, even thinking of good candidates was something of a problem.

At first we tried to get hold of the Science Citations database, an enormous network of scientific papers drawn from thousands of academic journals, linked to one another through their bibliographic citations. If I cite your paper, I am linked to you, and if my paper is cited by yours, you are linked to me. It wasn't quite what we were looking for (because papers usually only cite previously published papers, the links between them point only in one direction), but it was the best idea we had at the time. Unfortunately, the International Scientific Institute, which owned the database, wanted to make us pay for it, and we didn't have enough money.

Actually, they told us politely but firmly that if we gave them a single paper to use as a seed, for $500 they would send us a list of all the papers that cited that paper. For another $500, they would give us a list of the papers citing all of those papers, and so on. We thought this was absurd. If we had learned anything about networks, it was that as one searches out from some initial node (in this case, the seed paper), the number of nodes reached tends to grow exponentially. So for the first $500 the institute would have to deliver only a handful of papers, whereas for the third or fourth $500 they would have to find hundreds or thousands of times as many—for the same price! We briefly toyed with the idea of expending a couple of thousand dollars of Steve's precious research funds just to prove this to them, but reason eventually prevailed and we went back to thinking of other networks.

Our next attempt was more successful. In early 1997 a new game, called the Kevin Bacon Game, was born into popular culture, and it suited our interests perfectly. This game was invented by a group of fraternity brothers at William and Mary College who were apparently movie buffs of some note and who (in a state of altered reality no doubt) had come to the conclusion that Kevin Bacon was the true center of the movie universe. Just in case you haven't heard of it, here's how it works. The movie network consists of actors who are connected by virtue of having acted together in one or more feature films. We are not just talking about Hollywood here, but any movie made anywhere, anytime at all. According to the Internet Movie Database (IMDB), between the years 1898 and 2000, roughly half a million people have acted in over two hundred thousand feature films.

If you have acted in a movie with Kevin Bacon, you have a *Bacon number* of one (Bacon himself has a Bacon number of zero). Since Kevin Bacon has acted in quite a lot of movies (over fifty at the time of writing) and at last count had acted with 1,550 people, it follows that 1,550 actors have a Bacon number of one. This might sound like a lot, and certainly Bacon has acted with many more people than the average (which is only about sixty), but it is still less than 1 percent of the total population of movie actors. Moving outward from Bacon, if you haven't ever acted with him, but you have acted with somebody else who has, then you have a Bacon number of two. For example, Marilyn Monroe was in *Niagara* (1953) with George Ives, and George Ives was in *Stir of Echoes* (1999) with Kevin Bacon, so Marilyn has a Bacon number of two. In general, the object of the game is to determine an actor's Bacon number by figuring out his or her shortest path to the great man.

In Table 3.1 you can see what is called the *distance degree distribution* for the actor network, using Bacon as the origin. Almost 90 percent of all actors in the database have a finite Bacon number, which is another way of saying that they can be connected to Bacon through some chain of intermediaries in the network. So one conclusion we can draw straight away is that the actor network has a giant component in much the same way that a random graph does, once its critical connectivity has been exceeded. Another immediately apparent fact is that the vast bulk of actors have surprisingly small Bacon numbers—almost everyone in the giant component can be reached in four steps or less.

TABLE 3.1   DISTRUBUTION OF ACTORS ACCORDING
TO BACON NUMBER

BACON NUMBER	NUMBER OF ACTORS	CUMULATIVE TOTAL NUMBER OF ACTORS
0	1	1
1	1,550	1,551
2	121,661	123,212
3	310,365	433,577
4	71,516	504,733
5	5,314	510,047
6	652	510,699
7	90	510,789
8	38	510,827
9	1	510,828
10	1	510,829

One might also conclude, as did the fraternity brothers, that there is something special about Mr. Bacon, that he is somehow the fulcrum on which the universe of actors tips. But thinking about it for a moment longer, an entirely different interpretation seems more plausible. If it is true that Bacon can be connected to almost anyone in only a few steps, might it not also be true that *anyone* can be connected to anyone else in roughly the same number of steps? So, instead of calculating every-body's Bacon number, one might instead calculate *Connery numbers* or *Eastwood numbers*, or even *Pohlmann numbers* (Eric Pohlmann was an obscure Austrian actor who lived from 1913 to 1979, and acted in 103 films including *Return of the Pink Panther* and *From Russia with Love*). Going one step further and averaging over all possible starting points (that is, starting independently with every single actor in the giant component), one could obtain precisely the average path length that we had measured for our model networks.

All we needed was the network data. This, it turned out, wasn't a problem. Right about that time, Brett Tjaden and Glenn Wasson, both computer scientists at the University of Virginia, had launched a new Web site called the Oracle of Kevin Bacon that was quickly becoming one of the most popular destinations on the Web. Movie fans could type in the name of their favorite actor and the Oracle would instantly spit out the actor's path, just as we did for Marilyn Monroe. Figuring that in order to perform such computations, Tjaden and Wasson must have the network stored somewhere convenient, we wrote to Tjaden asking if we could have it. Somewhat to our surprise, he agreed imme-diately and even coached me through some of the idiosyncrasies of the raw data. Not long thereafter, we had calculated the average path length and clustering coefficient for the giant component, which at the time comprised roughly 225,000 actors. The result was clear, as we can see in Table 3.2. *In a world consisting of hundreds of thousands of individ-uals, every actor could be connected to every other actor in an average of less than four steps.* In addition, any actor's costars were very likely (80 percent of the time) to have starred with each other. No doubt about it— it was a small-world network.

Encouraged by this result, Steve and I immediately set about trying to find other examples. And because we wanted to test the generality of our models, we deliberately went looking for networks that had as little to do with social networks as possible. Thanks to the generosity of some colleagues of ours in the Electrical Engineering Department—Jim Thorp and Koenyi Bae—whose research was concerned with the dynamics of large power transmission systems, we were soon on our way. Steve and Jim were friendly with each other, and so we set up an appointment to talk to Jim and Koenyi about any network data they might have. As it

TABLE 3.2   STATISTICS OF SMALL WORLD NETWORKS

	L_{ACTUAL}	L_{RANDOM}	C_{ACTUAL}	C_{RANDOM}
MOVIE ACTORS	3.65	2.99	0.79	0.00027
POWER GRID	18.7	12.4	0.080	0.005
C. ELEGANS	2.65	2.25	0.28	0.05

L=Path Length; C=Clustering Coefficient.

turned out, they had a lot. In particular, they had a complete electronic map of the very same transmission network whose catastrophic breakdown in August 1996 is described in chapter 1. We immediately sat up and took notice, and pretty soon Koenyi was helping me sort out the somewhat labyrinthine notation that the Western Systems Coordinating Council uses to document its grid. After a few days of fiddling with the data, we had it in the right format and could run our algorithms on it. Much to our delight, we found exactly the same phenomenon as before. As Table 3.2 shows, the path length is close to that of a random network with the same number of nodes and links, but the clustering coefficient is much greater—just as our small-world models had suggested.

In an attempt to push our predictions a bit further still, the final network we considered was completely different again. We really wanted to find a neural network to calculate our statistics for, but we quickly discovered that neural data, like social network data, are painfully scarce. Fortunately, in all those years spent thinking about biological oscillators, Steve actually learned some biology, and after a few false starts he suggested that we look into an organism called

Caenorhabditis elegans, or *C. elegans* for short. It was, he told me, one of the model organisms that biologists have picked out for extensive study, and possibly someone had looked at its neural network.

Possibly! After only a cursory amount of research, and helped out by a biologist friend of Steve's who happened to be an expert on *C. elegans,* I quickly discovered that *C. elegans* is no bit player in the world of biomedical research. Alongside the fruit fly *Drosophila,* the bacterium *E. coli,* and possibly yeast, the tiny earth-dwelling nematode *C. elegans* is the most studied and, at least among worm biologists, the most celebrated of organisms. First proposed as a model organism in 1965 by Sydney Brenner, a contemporary of Watson and Crick, and thirty years later a pivotal player in the human genome project, *C. elegans* has spent over three decades under the microscope. Literally thousands of scientists have sought to learn not just something but *everything* about it. They aren't there yet, but their record of achievement is stunning, especially to someone stumbling across it for the first time. For instance, they have sequenced its entire genome, an achievement that might seem trivial in the shadow of the Human Genome Project but that having been done sooner and with vastly fewer resources, is probably equally impressive in its own way. They have also mapped out every cell in its body at every stage of its development, including its neural network.

One of the nice things about *C. elegans* is that unlike humans, the variation between specimens, even at the organismal level, is remarkably insignificant. So it is possible to speak of a typical neural network for *C. elegans* in a way that will never be possible for humans. Even more convenient, not only had a group of researchers completed the truly monumental task of mapping out how almost every neuron in its one-millimeter-long body is connected to every other, but also a second group had subsequently transcribed the resulting network data into a computer-readable format. Ironically, after two such spectacular scientific achievements, the end result resided exclusively in a pair of 4.5-inch floppy disks stuck inside the back cover of a book that was in the Cornell library. Or rather the book was there, but the librarian informed me that the floppies had been lost. Despondent, I returned to my office to think of another example network, but a couple of days later I received a call from the triumphant librarian—she had located the disks after all. Apparently no one was very interested in them, as I was the first person ever to check them out. Having obtained the disks and a computer sufficiently ancient that it had both a 4.5- and a 3-inch drive, the rest of the process was relatively simple. As with the power grid, the data required some finagling, but without too much trouble we were able to convert it into our standard format. The results, coming almost instantly this time, did not disappoint us—as Table 3.2 shows, the neural network of *C. elegans* was also a small world.

So now we had three examples, and finally some empirical valida-tion of our toy models. Not only did all three networks satisfy the small-world condition that we were looking for, but they did so despite huge differences in size, density, and most importantly, their underly-ing nature. There is nothing similar at all about the details of power grids and neural networks. There is nothing similar at all about the detailed way in which movie actors choose projects, and engineers build transmission lines. Yet at some level, in some abstract way, there is something similar about all of these systems because all of them are small-world networks. Since 1997, other researchers have started look-ing for small-world networks as well. As predicted, they are showing up all over the place, including in the structure of the World Wide Web, the metabolic network of *E. coli*, the ownership relations linking large German banks and corporations, the network of interlocking boards of directors of U.S. Fortune 1000 companies, and the collabora-tion networks of scientists. None of these networks are exactly social networks, but some, like the collaboration networks, are reasonable proxies. Others, like the Web and the ownership network, although not social in any real sense, are at least socially organized. And some of the networks considered have absolutely no social content at all.

So the models were right. The small-world phenomenon doesn't necessarily depend on the characteristics of human social networks, even the stylized version of human interactions that we had tried to build into the alpha model. It turns out to be something much more universal. Any network can be a small-world network so long as it has some way of embodying order and yet retains some small amount of disorder. The origin of the order might be social, like the interlocking patterns of friendships in a social network, or physical, like the geographical proximity of power stations—it doesn't matter. All that is required is some mechanism by which two nodes that are connected to a common third node are more likely to be connected to each other than two nodes picked at random. This is a particularly nice way to embody local order, because it can be observed and measured just by looking at the network data and it does not require us to know any details about the elements of the networks, the relationships between them, or why they do what they do. As long as A "knowing" B and A knowing C implies that B and C are, in turn, more likely to know each other than two elements picked at random, then we have local order.

But many real networks, especially those that have evolved in the absence of a centralized design, possess at least some disorder. Individuals in a social network exercise their agency, making choices about their lives and friends that cannot easily be reduced to their mere social context and history. Neurons in a nervous system grow blindly, subject to physical and chemical forces but without reason or design. For economic or political reasons, power companies build transmission lines that were not planned in earlier generations of the grid and that often cut across large distances and difficult terrain. Even institutional networks like the boards of directors of large corporations, or the ownership patterns linking the financial and commercial worlds—networks that one might expect to be ordered according to the Machiavellian designs of their creators—display the signature of randomness, if only because so many conflicting interests can never be reconciled in a coordinated fashion.

Order and randomness. Structure and agency. Strategy and caprice. These are the essential counterpoints of real networked systems, each inextricably intertwined with the other, driving the system through their endless conflict to an uneasy yet necessary truce. If our past had no influence on our present, if the present were irrelevant to the future, we would be lost, bereft not only of direction but also of any sense of self. It is through our surrounding structure that we order and make sense of the world. Yet too much structure, too strong a hold of the past on the future, can also be a bad thing, leading to stagnation and isolation. Variety is indeed the spice of life, for only with variety can order generate something rich and interesting.

And that is the real point behind the small-world phenomenon. Although we came to it by thinking about friendships, and although we shall continue to interpret many features of real networks in terms of social ties, the phenomenon itself is not restricted to the complex world of social relations. It arises, in fact, in a huge variety of naturally evolved systems, from biology to economics. In part it is so general because it is so simple. But it is not as simple as a mere lattice with a few random links added to it. Rather, it is the necessary consequence of a compromise that nature has struck with itself—between the stern voice of order and its ever subversive, unruly sibling, randomness.

Intellectually, small-world networks are also a compromise between the very different approaches to the study of networked systems developed over the decades in mathematics, sociology, and physics. On the one hand, without the perspective of physics and mathematics to guide us in thinking about global emergence from local interactions, we would never have tried to abstract the relationships embodied by the network beyond that of social relations, and we would never have seen the deep resemblance between so many different kinds of systems. On the other hand, without sociology to stimulate us, and without the insistence of social reality that real networks live somewhere between the cold order of lattices and the unconstrained disorder of random graphs, we would never have thought to ask the question in the first place.

CHAPTER FOUR

Beyond the Small World

• • • • • •

As HELPFUL AS IT WAS, OUR FOCUS ON SOCIAL NETWORKS DID lead us down the garden path a bit, and one of the most striking features of many real networks, including one of the very networks that Steve and I had in our possession, turned out to be something we never even thought to look for. One weekend in April 1999, I was sitting in my office at the Santa Fe Institute, where I was completing a postdoctoral fellowship, when I received a friendly e-mail message from László Barabási, a physicist at the University of Notre Dame, requesting the data sets from a paper on small-world networks we had published the year before. At the time, I had no idea what Barabási and his student Réka Albert were up to, but was happy enough to hand over the networks I had on hand and directed him to Brett Tjaden for the movie actor data. I should have paid more attention, because a scant few months later, Barabási and Albert published their ground-breaking paper in the journal *Science* establishing a whole new set of questions about networks.

What had we missed? Because our motivation derived from the small-world phenomenon, Steve and I were relatively unconcerned with the number of neighbors individuals typically have in their networks. We knew that sociologists had had a devil of a time measuring how many friends people have, and furthermore that whatever numbers their subjects came up with would depend a great deal on precisely how they understood the term *friendship* in the first place. Clearly, if *friend* implies "know on a first-name basis," one would get a completely different result than if it is taken to mean "would discuss personal issues with" or "would lend your car to for a week." As a result, we simply threw the problem into the too-hard basket and stopped thinking about it altogether. In so doing, however, we also made an assumption about the *distribution* of ties in our networks. Imagine that we could ask every person in a large friendship network how many friends they had (assuming some well-specified definition) and that they would all give us the correct answer. How many people would have only one friend? How many would have one hundred friends? How many would have no friends at all? In general, we could use our data to draw, as in Figure 4.1, what is called the *degree distribution* for the network. The degree distribution tells us, in a single picture, the probability that a randomly chosen member of the population will have a particular number of friends, or *degree* (not to be confused with degrees of separation).

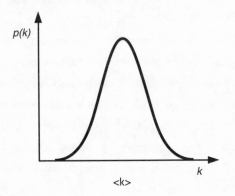

Figure 4.1. The normal distribution specifies the probability, *p(k)*, that a randomly selected node will have *k* neighbors. The average degree <k> lies at the peak of the distribution.

The assumption that Steve and I had made about the networks we were studying was that all such networks would have a degree distribution that looked roughly like the one in Figure 4.1. That is, not only would they exhibit a well-defined *average degree*, indicated by the sharp peak, but also most nodes would have degrees that were not too different from the average. Another way to say this is that the distribution tapers off, or *decays*, extremely rapidly on either side of the average, so rapidly, in fact, that the probability of anyone having many more friends than average is negligible, even in a very large network. In general, this is quite a sensible assumption. Very many distributions in the real world have exactly this property—so many that it is usually referred to as the *normal distribution*. For our purposes, normal-type degree distributions seemed like a reasonable guess for what real-world degree distributions should look like. They also satisfied another requirement of ours, that no one in the network be connected to any more than a small fraction of the entire population.

Remember that we were interested in the small-world phenomenon. Clearly, if some members of the population were connected to almost everyone, then the network would be small in an entirely trivial way. Think about the airline network. If you are flying somewhere, anywhere, even from a small airport, the first thing you do is fly to a major hub. From there you either fly directly to your destination or to another hub (unless, of course, the first hub itself is your destination). Even if you are flying from one small town to another small town on the other side of the planet, the total number of stopovers you need to make is rarely more than two or three, for the simple reason that each hub is connected to so many other airports, including many other hubs. Because we didn't think that social networks work this way—because no one can possibly know any more than a tiny fraction of the world's six billion people—we deliberately confined ourselves to networks with normal degree distributions to see how the world could be small *even in the absence* of hubs.

All this was entirely plausible and reasonable, but we made one big mistake. *We didn't check!* We were so convinced that nonnormal degree distributions weren't relevant that we never thought to look at which networks actually *had* normal degree distributions and which ones did not. We had the data sitting there staring at us for almost two years, and it would have taken all of half an hour to check it, but we never did.

SCALE-FREE NETWORKS

BARABÁSI AND ALBERT, MEANWHILE, WERE COMING AT MUCH THE same problem as Steve and I, but from a completely different angle. A native Hungarian, Barabási had been schooled in the Hungarian tradition of graph theory, including the Erdős model of random graphs. But as a physicist, he was dissatisfied with some of the more stringent requirements of random graph models and wondered what secrets might lie undiscovered in the huge quantities of real network data that were rapidly becoming available. One of the primary features of a random graph is that its degree distribution always has a particular mathematical form, named in honor of Siméon-Denis Poisson, the nineteenth-century French mathematician who studied the class of random processes by which it is generated. The *Poisson distribution* is not quite the same as the normal distribution, but they are similar enough that their differences need not concern us here. Essentially what Barabási and Albert did was to show that many networks in the real world have degree distributions that don't look anything like a Poisson distribution. Instead, they follow what is known as a *power law*.

Power laws are another very widespread kind of distribution in natural systems, although their origin is a good deal murkier than the origins of normal-type distributions like Poisson's. Power laws have two features that make them strikingly different from normal distributions. First, unlike a normal distribution, a power law doesn't have a peak at its average value. Rather, like Figure 4.2, it starts at its maximum value and then decreases relentlessly all the way to infinity. Second, the rate at which the power law decays is much slower than the decay rate for a normal distribution, implying a much greater likelihood of extreme events. Compare, for example, the distribution of sizes of people in a large population to the distribution of sizes of cities. The average height for an American adult male is roughly five feet nine inches, and although there are plenty of men who are shorter or taller than this, no one is even close to being twice as tall (almost twelve feet!) or half as tall (less than three feet). By contrast, the population of New York City, at just over eight million people, is almost 300 times the size of a town like Ithaca. Extreme differences like this would be inconceivable in a normal distribution but are entirely routine for power laws.

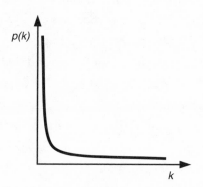

Figure 4.2.

A power-law distribution. Although it decreases rapidly with k, it does so much slower than the normal distribution in figure 4.1, implying than large values of k are more likely.

The distribution of wealth in the United States, for instance, resembles a power law. The nineteenth-century Parisian engineer Vilfredo Pareto was the first person to note this phenomenon, subsequently called *Pareto's law*, and demonstrated that it held true in every European country for which the relevant statistics existed. The law's main consequence is that very many people possess relatively little wealth while a very small minority are extremely wealthy. Because they are so highly skewed, the average properties of power-law distributions can be quite misleading. For instance, it is not very meaningful to talk about the average wealth of the U.S. population. Because it is so heavily dominated by the wealth of a few superrich individuals, who sit way out in the tail of the distribution, the actual average is considerably higher than what we would recognize as the wealth of a typical American. In the same way, a few superconnected nodes in a network can have an influence that is disproportionate to their number.

The key characteristic of a power-law distribution is a quantity called the *exponent*, which in essence describes how the distribution changes as a function of the underlying variable. For example, if the number of cities of a given size decreases in inverse proportion to the size, then we say the distribution has an exponent of one. In that case, we would expect to see cities the size of Ithaca roughly three times as frequently as cities like Albany (the capital of New York State) that are roughly three times as large, and ten times more often than cities like Buffalo that are ten times as large. But if instead the distribution decreases inversely with the *square* of the size, then we would say it has an exponent of two, and would expect that towns like Ithaca would arise nine times as often as towns like Albany and one hundred times as often as cities like Buffalo.

Rather than plotting the probability of an event as a function of its size (as in Figure 4.2), the easiest way to determine the exponent of a power law is to plot the *logarithm* of the probability versus the *logarithm* of the size. Conveniently, in this form (called a *log-log plot*), a pure power-law distribution will always be a straight line, just like in Figure 4.3. The exponent then is revealed simply as the slope of this straight

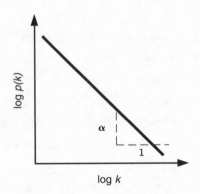

Figure 4.3. A power-law distribution on a log-log plot. The exponent alpha (α) is the slope of the line (the line drops by α for each unit on the horizontal axis).

line. So once we have enough data, all we need to do is plot it on a log-log scale and measure the slope of the resulting line. Pareto, for example, showed that regardless of which country he looked at, the wealth distribution was a power law with a slope somewhere between two and three where the lower the exponent, the greater the inequality. By contrast, if we plot a Poisson or a normal distribution on the same log-log scale, we see, as in Figure 4.4, that at some point it starts to curve down rapidly, displaying what is called a *cutoff*. In general, the cutoff sets an upper bound on whatever quantity the distribution represents. When it is applied specifically to the degree distribution of a network, the significance of the cutoff is that it limits how well connected any member of the population can be. If an average person can be connected to only a small fraction of the entire population, then the same will be true even of the best-connected person.

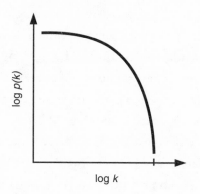

Figure 4.4. A normal-type distribution on a log-log plot. The *cutoff* occurs where the curve disappears into the horizontal axis.

Another way to think about the cutoff is that it defines an intrinsic *scale* for the distribution. And because a power law stretches on and on without ever encountering a cutoff, we say that it is *scale-free*. *Scale-free networks*, therefore, have the property—in striking contrast to a garden-variety random graph—that most nodes will be relatively poorly connected, while a select minority of *hubs* will be very highly connected. After looking at a range of network data, Barabási and Albert came to the surprising conclusion that many real networks, including the movie actor network that Steve and I had examined, the physical network of the Internet, the virtual link structure of the World Wide Web, and the metabolic networks of several organisms, were scale-free. In light of several decades of assumptions to the contrary, this would have been a striking observation on its own. But what really grabbed the interest of the network community is that they went one step further, proposing a simple and elegant mechanism by which such networks could *evolve* over time.

THE RICH GET RICHER

IT TURNS OUT THAT THE ORIGIN OF THE POISSON DEGREE distribution in a random graph, and its corresponding cutoff, lies with its most basic premise: that links between nodes come into existence entirely independently of one another. At any point in the construction process, poorly connected nodes are just as likely to make or receive new connections as are the best-connected nodes. As one might expect in such an egalitarian system, things average out over time. An individual node can be unlucky for a while, but eventually it has to be on the receiving end of a new connection. Likewise, no run of pure luck can go on forever, and so even if one node gets picked more frequently than average for some period of time, eventually the others will catch up.

Real life, however, is often not that fair. Particularly in matters of wealth and success, the rich always seem to get richer, usually at the expense of the poor. This phenomenon has been with us a long time—at least as long as the Bible, in which Matthew observes, "For unto every one that hath shall be given, and he shall have abundance; but from him that hath not shall be taken away even that which he hath." In the context of networks, the *Matthew effect*, as it was coined by the great twentieth-century sociologist Robert Merton, equates to well-connected nodes being more likely to attract new links, while poorly connected nodes are disproportionately likely to remain poor.

Barabási and Albert proposed that a special case of the rich-get-richer effect drives the evolution of real networks. Specifically, if one node has twice as many links as another node, then it is precisely twice as likely to receive a new link. They also proposed that unlike standard random graph models, in which the number of nodes remains fixed and only links are added, a realistic network model should allow the population itself to grow in time. Barabási and Albert, therefore, started with a small group of nodes and then systematically added both nodes and links, where at each time step a new node is added and allowed to con-

nect to the existing network by extending a fixed number of ties. Each
node already in the network is eligible to be on the receiving end of each
tie, with a probability that is directly proportional to its current degree.
The oldest nodes in the network, therefore, have an advantage over more
recent additions. Because there are only a few of them initially, they are
likely to attract a few early connections, and then the rich-get-richer rule
tends to lock in that advantage for all time. The result, as Barabási and
Albert demonstrated, is that over a sufficiently long time horizon the
degree distribution of network converges to a power law distribution,
reminiscent of the distributions they had seen in their data.

Why is this important? To start with, a scale-free degree distribution is so different from a Poisson distribution that anyone attempting
to understand the structure of real networks cannot help but pay attention to it. Clearly the standard model of random graphs proposed by
Erdős and Rényi has some serious problems, not just because it fails to
predict the clustering that we discussed earlier, but also because it cannot explain why Barabási and Albert found the degree distributions
that they did. Simply recognizing that the world is strikingly different
from what has previously been assumed is a significant step forward.
The explanation of preferential attachment, however, makes an additional statement about the way the world works: that small differences
in ability or even purely random fluctuations can get locked in and lead
to very large inequalities over time. And finally, as we will see in later
chapters, scale-free networks also have a number of other properties,
such as their vulnerability to failures and attacks, that distinguish them
from normal networks and that are of significant practical interest.

Although they didn't know it at the time, Barabási and Albert were
not the first to propose a preferential growth model to account for the
existence of power-law distributions. As long ago as 1955, the polymath
and Nobel Prize–winner Herbert Simon (of bounded rationality fame)
devised an almost identical model to explain the size distribution of business firms. This particular distribution is an example of *Zipf's law*, named
after the Harvard linguistics professor George Kingsley Zipf, who in

1949 used it to describe a quite different distribution altogether: the rank-ordered frequency with which words occur in English language texts. (The word *the,* it turns out, is the most commonly used word, followed by *of,* and so on.) Zipf *ranked* all the words in a number of large volumes of text according to how often they occurred, and showed that when their frequencies were plotted against their rank, the resulting distribution was a power law. Zipf then went on to show that his law also applied (among other things) to the rank-ordered distribution of city sizes (where the exponent turns out to be close to one) and the assets of business firms.

Zipf himself attributed the phenomenon to what he called the "principle of least effort," an intriguing concept, but one that remained frustratungly elusive despite a (rather long) book that he wrote with that phrase in the title. Six years later, Simon and his collaborator, Yuji Ijiri, proposed a simple model that assumed—like Barabási and Albert's—that individual cities (or in Simon and Ijiri's case, businesses) grow in a more or less random fashion, but that their probability of growing by a given amount is proportional to their current size. Large cities like New York, therefore, are more likely to attract new arrivals than small cities like Ithaca, thereby amplifying initial differences in size and generating a power-law distribution in which a few "winners" account for a disproportionately large share of the overall population.

In reality, there is nothing random about New York being larger than Ithaca—New York lies at the mouth of one of the major rivers of the eastern seaboard, whereas Ithaca nestles in the midst of a sleepy farming community—but that wasn't the intention of Simon's model. He wouldn't have denied the importance of geography and history in determining which *particular* cities became major metropolises, anymore than Barabási and Albert would have denied that a promising business plan and access to venture capital were essential to becoming a highly visible, highly connected Web site. Rather the point is that once an individual city, business, or Web site becomes large, then *regardless of how it did so,* it is more likely than its smaller counterparts to grow larger still. The rich have many ways of getting richer, some deserved and others not, but as far as the resulting statistical distribution is concerned, the only thing that matters is that they do.

The sheer generality of Barabási and Albert's model promised a new way to understand the structure of networks as dynamically evolving systems. It didn't matter whether the networks were of people, Internet routers, Web pages, or genes. As long as the system obeyed the two basic principles of growth and preferential attachment, the resulting network would be scale-free. As Simon himself pointed out, however, even elegant and intuitively appealing models can be misleading. Sometimes the details do make a difference.

GETTING RICH CAN BE HARD

ONE PARTICULARLY PROBLEMATIC DETAIL OF SCALE-FREE NETWORKS is that power-law distributions are only ever truly scale-free when the network is infinitely large, whereas in practice, every network we encounter is finite. Finite size effects cause problems for just about every statistical technique, but they're particularly troubling for power laws because the finite size of the systems always forces a cutoff on the distribution. In more concrete terms, no node in any real network can ever be connected to more than the entire rest of the population. So even if the underlying probability distribution is scale-free, the observed distribution must display a cutoff somewhere, typically one that is far below the system size. Hence, the actual degree distributions that the scale-free network model was designed to explain really exhibited two regions, like in Figure 4.5: the scale-free region, which shows up as a straight line on the log-log plot, and a finite cutoff.

The confusion for the observer arises in deciding whether or not an observed cutoff is just a function of the finite system size or is actually due to a more fundamental property of the system. For example, the number of friends that people have is not limited by the size of the global population, which is plenty large enough for most people to have hundreds or even thousands of times more friends than they actually do. No, the real constraint is with people themselves, who only have enough time, energy, and interest to befriend so many others before the sheer

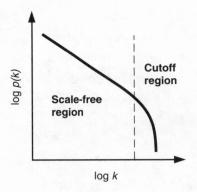

Figure 4.5. In practice, power-law distributions always display a characteristic cutoff because of the finite size of the system. The observed degree distribution, therefore, is only ever a straight line on a log-log plot, over some range.

effort of it all overwhelms them. Even if the Matthew effect applies to networks like the World Wide Web, it isn't clear that it should work the same way in all or even most networks. Worse still, sometimes the cutoff is so severe that the distribution in Figure 4.5 becomes hard to distinguish from distributions like that in Figure 4.4, which are not scale-free at all.

Some evidence that scale-free networks may not be as widespread as they first seemed appeared about a year after Barabási and Albert's original paper. A young physicist, Luis Amaral, along with a number of his colleagues, including H. Eugene Stanley—one of the giants of statistical physics (and also Barabási's former advisor)—published a paper in the *Proceedings of the National Academy of Sciences,* in which they looked at the degree distributions of a number of real networks. They showed that although some of them resembled power-law distributions (albeit with finite cutoffs), some clearly did not. Most striking was the social network of a community of Mormons in Utah, which looked far more like a plain old normal distribution than anything exotic. Further evidence of non-scale-free networks comes from what now seems like the distant past, in one of Anatol Rapoport's papers, where he considers the friendship network in a Michigan high school. Rapoport, like Steve and me, wasn't all that interested in the degree distribution, but he did at least take the time to plot his out, and although it is not the familiar Poisson distribution of a random graph, neither is it scale-free.

That the world is more complicated than a simple model like Barabási and Albert's should come as no surprise, nor should it detract from their achievement. The introduction of scale-free networks is one of the central ideas in the new science of networks, and it prompted a veritable flood of papers, particularly in the physics literature. The arrival of the physicists into the science of networks has brought the kind of mathematical and computational muscle that the field has lacked for so long, and as result, the past few years have been a tremendously creative and exciting time for those of us involved. But almost immediately it became apparent that mere muscle wasn't going to be enough. Just as our original small-world models had missed a number of features of the real world, so too did the simple principles of network growth and preferential attachment.

The essential limitation with the scale-free view of networks is that everything is assumed to come for free. Network ties in Barabási and Albert's model are treated as costless, so you can have as many of them as you are able to accumulate, without regard to the difficulty of making them or maintaining them. This assumption may indeed work for examples like the Web, but it is usually not true for human, biological, or even engineering systems like the power grid. Information also is assumed to be free, so a newly arrived node can find and connect to any node in the world, and the only relevant factor is how many connections each existing node currently maintains. In reality, however, new arrivals start in a particular part of a large system and need to learn about it through a costly process of search and discovery. When we move to a new city, we can't simply find the person who has the most friends. We may be more likely to meet someone who has a lot of friends than someone who has very few, but other factors play a role as well. And once we have made our initial contacts, the social structure in which we are now embedded renders some people far more accessible than others, regardless of how well connected they are on a grand scale.

This was precisely the effect we had tried to capture in our small-world models, and we remained convinced of its importance, but the scale-free models had no element of social structure in them whatsoever. On the other hand, Barabási and Albert's elegant results had convinced us that the tools available to study random networks were too powerful to ignore. Somehow we needed to harness the mathematics of the physicists to the problem of social structure, and in doing so break through the barriers that had beset Anatol Rapoport five decades earlier. Most of all, we needed a new idea.

REINTRODUCING GROUP STRUCTURE

ON FEBRUARY 20, 2000, A DATE I REMEMBER ONLY BECAUSE IT happened to be my birthday, Steve and I met at the annual meeting of the American Association for the Advancement of Science (AAAS), in Washington, D.C., to organize a session on networks and the history of the small-world problem. Also presenting at the session was the sociologist Harrison White, a man who has an interesting history of his own. White actually began his academic career as a theoretical physicist, studying solid-state physics at MIT in the early 1950s. Like many young physicists, then and now, he quickly realized that the big unsolved problems of mainstream physics had already been well defined, and everyone seemed to know what they were. Thousands of smart, hard-working, ambitious graduate students and postdocs just like him were slaving away in labs around the world, hoping to make the next big breakthrough. Unless you happened to be smarter than all of them, work harder, and somehow be lucky enough to have the right idea at the right time, your chances of success are, as we say in Australia, "Buckley's or none" (and at least according to the legend, Buckley didn't have any chance either). Every young physicist experiences this realization of hopelessness, so in that sense, Harrison was pretty normal. What made him unusual was what he chose to do about it.

In his first year of graduate school at MIT, Harrison had taken a course on nationalism with the political scientist Karl Deutsch and had found it captivating. Encouraged by Deutsch, he decided to abandon physics for the social sciences. Taking advantage of a one-year fellowship from the Ford Foundation, he went back to graduate school at Princeton and got another Ph.D., this time in sociology. But he would always remain in part a physicist. Decades before the word *interdisciplinary* had permeated university campuses and funding agencies alike, Harrison was the quintessential interdisciplinarian, a benign Trojan horse through which the ideas and techniques of contemporary physics could invade and reshape sociology. At Harvard in the 1970s, he was a colleague of Stanley Milgram, and even did some work on the small-world problem. But he also started and ran a program in applied mathematics that trained some of the most influential sociologists of the next generation, and made a number of seminal contributions to the modern theory of social networks. In his seventies now, Harrison is famous not only for his irascible manner and impenetrable writing but also for his profound generosity, astounding breadth of interest, and occasionally startling insight.

At the conference, Harrison was his usual Delphic self, but he said one thing that succeeded in jolting some old cogs into motion. The essence of his talk was that people know each other because of the things they do, or more generally the *contexts* they inhabit. Being a university professor is a context, as is being a naval officer. Flying frequently for business is a context. Teaching climbing is a context. Living in New York is a context. All the things we do, all the features that define us, and all the activities we pursue that lead us to meet and interact with each other are contexts. So the set of contexts in which each of us participates is an extremely important determinant of the network structure that we subsequently create.

Stimulated by Rapoport's work, I had been struggling for some time with the idea of building random networks that incorporated social structure in a way that was less messy than the alpha model Steve and I had worked on initially, but which didn't rely on artificial lattice substrates like the beta model. The problem was that as soon as we removed the lattice metric, we no longer had any way of determining how close anybody was to anybody else, and thus how likely they were to be connected. In random graphs, this isn't a problem because everyone is equally likely to be connected anyway, and in Barabási's scale-free networks, the connection probabilities depend only on the degree. But once we start introducing any kind of social or group structure, we need some basis for distinguishing "close" from "far away." In fact, without notions of close and far away it is no longer clear how one should even go about defining social structure in the first place. After all, what is a social group if not a set of individuals to whom you are in some sense closer than you are to the rest of the world?

It was the beginnings of a solution to this problem that dawned on me while listening to Harrison's talk. Instead of starting with a notion of distance and using it to construct the groups, why not start with the groups and use them to define a measure of distance? Imagine that instead of individuals in a population choosing each other directly, they simply choose to join a number of groups, or more generally, participate in a number of contexts. The more contexts two people share, the closer they are, and the more likely they are to be connected. Social beings, in other words, never actually start out on a tabula rasa in the same way that the nodes in our previous network models had done, because in real social networks, individuals possess *social identities.* By belonging to certain groups and playing certain roles, individuals acquire characteristics that make them more or less likely to interact with one another. Social identity, in other words, drives the creation of social networks.

As simple as it sounds, this view of networks was actually fundamentally different from the one with which we had been operating up to that point, because it required us to think simultaneously about two distinct kinds of structure—*social structure* and *network structure*—rather than just one. Of course, this view is entirely natural for sociologists. As we discussed earlier, sociologists have thought long and hard about the relation between social and network structure. But it is not at all natural for physicists or mathematicians, to whom the idea of a network node possessing an identity sounds vaguely ridiculous. Nevertheless, the intuition was so appealing, I was amazed that I hadn't thought of it before. Actually, I had thought of it before. It was, in fact, the very first idea I had ever proposed to Steve as a model of a social network, back when we first started thinking about the whole thing. But for a number of technical reasons, we hadn't been able to make it work, so we had dropped it and gone on to the conceptually easier lattice models. A few years later, it still seemed like a hard problem, but by this stage Steve and I had discovered our new secret weapon: Mark Newman.

Mark Newman is precisely the kind of person who makes you wonder why you ever bother trying to do anything. A brilliant physicist and maestro on the computer, Mark is also an accomplished jazz pianist, composer, singer, and dance instructor and even manages well on a snowboard. Still in his midthirties, he has written four books and published dozens of papers in physics and biology journals, has built a reputation as a good teacher, and has invented a number of original computer algorithms—all without working nights or weekends! More than anything, however, he is fast—unbelievably, indefatigably fast. Working with Mark is like stepping onboard an express train without checking first which line you're on—you're guaranteed to get *somewhere* very quickly, but you're too busy hanging onto your hat to figure out where until you arrive, usually exhausted. The train, meanwhile, is already off writing another paper.

Getting Mark interested in our problem took a bit of effort, but fortunately he and I had already written a few papers together in Santa Fe, exploring some of the mathematical properties of the beta model, and this had given him a taste for network-related problems. At my suggestion, Steve had also invited Mark to Cornell to give a talk, and they had taken an instant liking to each other, so the idea of a collaboration was appealing to all of us. The main difficulty was that by then, in early 2000, I was living in Cambridge, Massachusetts. I had moved there the previous fall to work with Andrew Lo, a finance economist at the MIT Sloan School of Management and an old friend of Steve's from their graduate school days at Harvard. Mark, meanwhile, was back in Santa Fe, and Steve was still in Ithaca, so our ideas had to be bounced around via e-mail, which wasn't proving very effective. Eventually, however, we found a long weekend in May when we could meet in Ithaca to talk about a new project. Steve neglected to mention that the weekend we had picked was also Cornell's commencement weekend, during which the campus and most of the town disappear under a teeming throng of parents, brothers, sisters, cousins, aunts, and even a few delirious students. Nevertheless, we managed to sequester ourselves in Steve's house in Cayuga Heights and get some serious work done. Or rather Mark got some serious work done, while Steve and I sat and watched in admiration as the machine wound into high gear.

AFFILIATION NETWORKS

THE TECHNICAL TRICK THAT HELPED US TO NAIL DOWN THE CONCEPT of distance as a function of group structure was to express it in terms of what is called an *affiliation network*. Like the actor network of chapter 3, in which two actors are considered connected if they have acted in the same movie, two nodes in an affiliation network can be said to be affiliated if they participate in the same *group* or, in Harrison's terminology, *context*. The affiliation network then becomes the *substrate* on which the actual network of social ties is enacted. Without any affiliations, the chance that two people will be connected is negligible; the more affiliations they have, and the stronger each affiliation is, the more likely they are to interact as friends, acquaintances, or business associates, depending on the nature of the contexts involved. But before we could even start on the problem of building social networks out of affiliation networks, we first had to understand the structure of the affiliation networks themselves, and it was this problem that Steve, Mark, and I elected to start on that weekend in Ithaca.

Affiliation networks are an important class of social networks to study in their own right, not only because affiliations form the basis for other kinds of social relations like friendships or business ties, but also because they arise in a wide range of non-social-network applications that are nevertheless economically and socially interesting. For example, when you go to Amazon.com to buy a book, and underneath your selection it lists "people who bought this book also bought . . .," that's an affiliation network. It consists of people on the one hand and books on the other. By buying a book, individuals become affiliated with anyone else who ever bought that same book, in effect choosing a new "group" to which they now belong. The network of movie actors is also an affiliation network consisting of actors on the one hand and movies on the other. By acting together in a movie, two actors are considered affiliated. A similar description can be applied to corporate directors who sit together on boards of companies, and scientists who write papers together. In fact, one of the earliest affiliation networks to receive any attention was the coauthorship network of mathematicians that includes Paul Erdős, the inventor of random graph theory who we encountered in chapter 2.

Another reason to study affiliation networks is that the data we have are unusually good because at least for contexts like membership in clubs, participation in business activities, and collaboration on joint projects like movies or scientific papers, it is particularly clear who belongs to what. Recently, a great deal of such data has become available electronically in the form of on-line databases, so even very large networks can be constructed and analyzed rapidly. Even better, in at least some cases like the Amazon.com example and some of the collaboration networks we will discuss later, the data are recorded automatically by the actual individuals, with consumers making decisions about what to purchase or researchers submitting scientific papers in real time. By distributing the effort of data entry to the members of the network themselves, rather than concentrating it in the hands of a database manager, the main limitation on data recording is virtually eliminated, and the resulting databases can essentially grow without bounds—a far cry from the collection and recording methods of even a decade ago.

Because affiliation networks always consist of two types of nodes— let us call them *actors* and *groups*—the best way to represent them is in terms of what is called either a *bipartite* or a *two-mode* network. As shown in the middle panel of Figure 4.6, in a bipartite network, the two sets of nodes are represented separately and only nodes of different types can be connected, via a relation that we can interpret as *belongs to* or *chooses*. Hence, actors are connected only to groups and groups only to actors. Whereas a *single-mode* (or *unipartite*) network, like the networks we have considered up to this point, is characterized by a single degree distribution, two-mode networks require two such distributions: the distribution of group sizes (how many actors belong to each group) and the distribution of how many groups each actor belongs to.

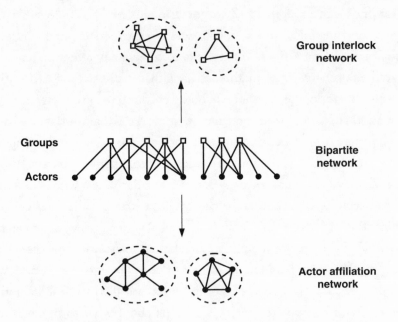

Figure 4.6. Affiliation networks are best represented as bipartite networks (center) in which actors and groups appear as distinct kinds of modes. Bipartite networks can always be projected onto one of two single-mode networks representing affiliations between the actors (bottom) or interlocks between groups (top).

As different as a bipartite network seems, it can always be represented as two single-mode networks, one consisting of the actors and the other consisting of the groups, by projecting it onto the two sets of nodes, as we do in Figure 4.6. The actor mode is the familiar one—two actors are connected (affiliated) if they both belong to at least one group. But groups can also be affiliated by virtue of common membership—if at least one actor belongs to two groups, we say that they *over-*

lap or *interlock.* The result of this projection trick is that in principle, a single bipartite network contains all the information relevant to both the *actor affiliation network* and the *group interlock network,* shown in Figure 4.6 at the bottom and top, respectively. What Steve, Mark, and I were hoping to do was understand the observable properties of the single-mode networks, in terms of the bipartite representation. The reason for doing it this way goes back to Harrison's discussion at the AAAS conference. The one-mode networks are representations of the relationships that one might actually observe in measurable network data, like the data that network analysts normally collect—a list, effectively, of who knows who. But what this kind of network data can't tell us is where these relationships come from.

As we discussed in chapter 2, traditional network analysis attempted to avoid this problem by devising techniques to extract group structure from the network structure alone. In terms of Figure 4.6, this would amount to re-creating the group interlock network (at top) solely from one's knowledge of the actor affiliation network (bottom); that is, without knowing the bipartite graph in the middle. But as you can also see from Figure 4.6, for even a relatively small bipartite network, the group projection (top) is not much less complicated than the actor projection (bottom). So not only is the relationship between the two networks difficult to extract without knowing where they both come from, but it's not exactly clear how the exercise would clarify matters anyway. By starting with an explicit representation of the social structure—that is, the full bipartite representation—we hoped to understand the structure of both the affliation and interlock networks simultaneously.

DIRECTORS AND SCIENTISTS

AT ABOUT THE TIME THAT THE THREE OF US MET UP IN ITHACA, I received an e-mail from Jerry Davis, a professor in the business school at the University of Michigan, asking for computational assistance with some network data that he and his collaborator, Wayne Baker, were studying. For some years, Davis had been deeply interested in the social structure of corporate America, in particular, the interlocking structure of corporate boards of directors. This is not a trivial social network. Roughly eight thousand directors sit on the boards of the U.S. Fortune 1000 companies, and this relatively small number of individuals, along with their executive officers, play a critical role in determining the economic landscape of the country and, to a lesser extent, the world. Because most of the players in this game are accountable only to their shareholders (if that!), and the maximization of corporate wealth is not necessarily in the best interests of the general population, the environment, or enlightened governance, an important question is whether or not the far-flung elements of the corporate world are capable of coordinated action, in violation of the assumed principles of market competition. In the aftermath of widespread accounting scandals in the energy-trading and telecommunications industries, it seems more important now than ever to identify potential mechanisms for corporate collusion.

Economists historically have not thought much about this issue because they generally assume that markets always govern the interactions between firms. But sociologists like Davis think about it a lot. Somebody sitting on the boards of two different firms will naturally form a conduit for information flow between them and is likely also to push their interests into alignment. There are rules about joint membership, of course—one cannot sit on the boards of different companies that are direct competitors, for instance—but mutual interests are often more subtle than any rules can track. Also, corporate coordination is not always a bad thing. If the U.S. corporate sector as a whole is to respond nimbly and efficiently to a rapidly changing global economic environment, it helps for corporate officials to talk to each other in settings more intimate than the business pages.

Naturally corporate executives and directors have multiple forums for interaction, both formal and informal, of which boardrooms are but one. But since the boardroom is where most major changes of corporate strategy are either conceived or approved, it seems a particularly significant context to study. Furthermore, unlike the informal interactions between CEOs on the golf course or over lunch at the China Grill, the membership of corporate boards is publicly available data, and thus lends itself to analysis. Davis and Baker wanted to know if their corporate board network was a small world in the sense that it was highly clustered, but any two directors could still be connected by only a short number of intermediaries. It didn't take very long to verify that it was, so we could add another specimen to the growing list of known small-world networks. But this fact no longer seemed surprising to us, and so I asked Davis if he would mind us doing some more detailed analysis of their data. He generously agreed.

In the meantime, Mark had been doing some homework on his own. In the mid-1990s, Paul Ginsparg and Geoffrey West, two physicists from Los Alamos National Laboratory, had started a minor revolution in scientific publishing by creating an on-line, electronic repository for prepublication research papers spanning the various subdisciplines of physics. The physics community, as frustrated as anyone by the traditional journal-based publication process and ever eager to catch the next wave, flocked to the new outlet, known as the *LANL e-print archive*. The archive serves at least two functions that make it an innovative scientific institution. First, it provides researchers with a practically instantaneous publication option that requires them only to upload their paper onto the archive's servers. Second, it provides the rest of the research community with equally rapid access to everybody else's work, thus dramatically speeding up the cycle of ideas and innovation. Whether or not this almost unrestricted ability to publish one's own work is entirely a good thing for the progress of science is yet to be seen. But most physicists apparently think it is, at least if their enthusiasm for uploading and downloading papers is anything to go by.

Above and beyond its institutional significance, the archive also serves as an object of scientific investigation, as a network of collaborations between scientists. In the half decade since its establishment, roughly one hundred thousand papers have been posted to the archive by over fifty thousand authors. Obviously these numbers are only a fraction of the total population of physicists and their papers over all time, but they are significant enough to be representative at least of the contemporary social structure of the discipline. Through Ginsparg, Mark had managed to get the entire database of papers and authors, from which he could reconstruct the corresponding collaboration network as a bipartite graph.

Never one to do things by halves, Mark had also managed to get his hands on some even more impressive data, namely, the MEDLINE database of biomedical researchers and papers, compiled over a considerably longer period of time than the e-print archive and consisting of over 2 million papers and 1.5 million authors. These numbers are absolutely off the charts for social network analysis (Davis's data set is considered large, and it's in the thousands). Mark not only had to use the giant, new Intel cluster that was just being installed at the Santa Fe Institute to do the calculations, but also had to improve on some standard network algorithms in order not to tie up even that machine for the next few years. And as if that wasn't enough, Mark also acquired two smaller (but still enormous by social network standards) databases from the high-energy physics and computer science communities.

From an economic perspective, a network of scientific collaborators is not as obviously significant as a network of corporate directors. But viewed over a longer time horizon, the ability of the scientific community to innovate, and also to agree, has profound (if somewhat indeterminate) consequences for the production of new knowledge and its conversion to technology and policy. Inasmuch as the social structure of collaborations is a mechanism for scientists to learn new techniques, dream up new ideas, and solve problems they would not have been able to solve alone, then it is critical to the healthy functioning of the scientific enterprise. In partucular, one would hope that even a very large collaboration network of scientists would be connected as a single community and not many isolated subcommunities.

So by the time we met in Ithaca that weekend in May, we had not only some theoretical ideas about affiliation networks but also a pretty good picture of what sort of empirical phenomena our models would have to explain. One of the more striking features of the collaboration networks, for example, was that the majority of authors in each of the networks were indeed connected in a single component, within which any working scientist could be linked to any other through only a short chain of collaborators (typically four or five). We were not exactly surprised by these findings, as we had already observed this property in the movie actor network. However, remember that some of Mark's data sets had been created in only five short years and already consisted of tens of thousands of authors, so the scientists (who tend to be fairly narrowly focused anyway) had not had as long as the actors to get connected. Furthermore, the typical number of authors on any given paper is only about three, much less than the size of an average movie cast (about sixty), making it less obvious that everyone should be so well connected.

Nevertheless, this phenomenon is something that random graph theory can explain rather easily. In random graphs, as in the alpha model of chapter 3, it is not possible to have two large components of roughly equal size that are not themselves connected. The reason is simply that if two such components *did* exist, then almost inevitably one member of one of the components would eventually, by random chance, connect to one member of the other component, at which point they would cease to be separate. The surprise perhaps is that this result appears to hold even in nonrandom networks, where forces like disciplinary specialization tend to segregate communities. But as we saw with the alpha model, even the tiniest amount of randomness seems to do the trick. The presence of high connectivity and short global path lengths therefore augured well for a random network model.

COMPLICATIONS

STUDYING THE DATA A BIT CLOSER, HOWEVER, QUICKLY REVEALED a number of features that looked nothing like a random network at all. For starters, the collaboration networks were all very highly clustered in the by-now-familiar manner of small-world networks. And second, the distributions of how many papers each author had written, and the number of coauthors with whom each had written them, looked more like Barabási and Albert's power-law distribution than the sharply peaked Poisson distributions that are the hallmarks of random graphs.

Matters got even more complicated once we started looking at Davis's data on boards of directors. *Everyone* in the entire network of directors, not just a large fraction of them, was connected, and the corresponding degree distributions were nothing like either scale-free or normal random networks. Sitting on the board of a Fortune 1000 company is not a trivial undertaking, so not surprisingly most directors— nearly 80 percent of them in fact—belong to only one board. The distribution then falls off exponentially fast, much faster than a power-law distribution but slower than a Poisson or normal distribution. Incidentally, the most connected director in the network is none other than Vernon Jordan, former President Bill Clinton's best friend who gained considerable notoriety during the Monica Lewinsky scandal. (Revlon, at which he supposedly offered her a job, is one of his nine boards.) Meanwhile, the distribution of codirectors—how many other directors each director sits on a board with—was just plain weird. As Figure 4.7 shows, it exhibits not one but two distinct peaks and then a long tail that doesn't even appear to decay smoothly. No standard distribution listed in any statistics book will fit this stretched, lumpy mess. So what kind of network was this anyway? And was there any way to understand distributions like this in terms of a theory that also explained the structure of the collaboration networks?

The key, as suggested earlier, was to model affiliation networks in terms of the full, bipartite (two-mode) representation in Figure 4.6—that is, by treating actors and groups as distinct kinds of nodes, allowing actors to connect only to groups and vice versa. Starting with the properties of the bipartite version, we could then compute the expected properties of the corresponding single-mode projections (the top and bottom diagrams of Figure 4.6). To move beyond mere description, however, we had to make some assumptions, and it made sense to start simple. Taking the two distributions of the bipartite network (groups per actor and actors per group) as given, we assumed that the matching between actors and groups occurred in a more or less random fashion. Clearly this isn't the case in the real world, where decisions about which groups to join are generally planned and often quite strategic. But as we had done so often

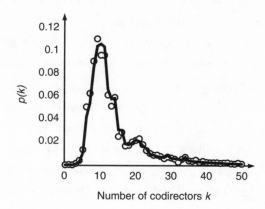

Figure 4.7. Distribution of codirectors for Jerry Davis's boards of directors data. The circles are the real data, and the line is the theoretical prediction.

in our models before, we hoped that the decisions of individual actors were sufficiently complicated and unpredictable that it would not be possible to distinguish them from simple randomness.

Utilizing a powerful mathematical technique for studying the properties of random distributions, Mark, Steve, and I showed that most of the classic properties of random single-mode networks (the ones studied previously, and far more formally, by Erdős and Rényi) extended very naturally to the two-mode version. The properties that we had observed in the scientific collaboration networks, like short path lengths and the existence of a giant component, all followed in a straightforward manner from the assumption that actors choose groups at random. Even more interesting and far less expected, our model also accounted for the bizarre degree distribution of Davis's data (as Figure 4.7 shows, the agreement between the theory and the data is so close, it's uncanny) and almost all of its clustering.

But didn't we show earlier that random networks don't have any clustering? Well yes, but this is what makes the bipartite representation of affiliation networks so useful. Because, by definition, every actor in a group is affiliated with every other actor in that group, in the one-mode projection of a bipartite network, every group is represented as a fully connected *clique* of actors. Affiliation networks (like the bottom panel of Figure 4.6) are thus networks of overlapping cliques, locked together via the comembership of individuals in multiple groups. Because this feature is a property of the *representation* of the network, and not of any particular matching procedure, it is true regardless of how individuals and groups are matched. Even a random bipartite network—one that has no particular structure *built into it* at all—will be highly clustered. On the other hand, the randomness causes the networks to remain highly connected and to display short global path lengths. In other words, random affiliation networks will always be small-world networks!

This was a particularly encouraging result, not so much because we needed yet another way to generate small-world networks (that much was easy), but because the small-world properties arose in such a natural fashion. Simply by representing the problem in what seemed a sociologically plausible manner—by assuming, in effect, that people know each other because of the groups and activities in which they participate—we were able to generate many features of real social networks. Not surprisingly, our models still made a number of simplifying assumptions, most significantly that actors choose groups at random. But these shortcomings are not only rectifiable; they actually demonstrate how strong the results are. If even the simplest possible mechanism for actors choosing groups can generate at least plausible network structures, then the basic approach seems to be the right one.

There is lot of work still to be done, however, and once again dynamics seems to be the key. People may know who they know because of what they do, but people also try new things because of who they know. Your friends invite you to parties or drag you along on their favorite activities. Your colleagues involve you in new projects or suggest contacts who might be able to help you with a problem. And bosses suggest new opportunities within the firm or even outside. It is through your current social contacts, in other words, that you often get the information to expand your horizons, thereby altering the social structure within which you move and generating the next round of acquaintances with whom you will share it. The true power of the bipartite approach is that all these processes—the dynamics *of* the network—can be represented simply and explicitly in a single framework, one that can track the evolution of both the social and the network structure, and the endless folding of one into the other that constitutes the heart of the social process.

But what does all this mean? Even if we understand how people end up creating a network structure out of a social structure (and vice versa), what can they do with it once it's there? And in turn, by constraining the information people have available to them and by exposing them to influences that may lie beyond their ability to control, what effect can the network have on the people in it? As we mentioned in chapter 1, the answers to these questions depend very much on the sort of action or influence—dynamics *on* the network—in which one is interested. Different kinds of dynamics on networks, therefore, have to be explored in different ways, sometimes leading us to new insights about the networks themselves. In order to get a handle on the matter, we will need to return once more to Stanley Milgram's small-world problem, which has turned out to be a good deal more subtle than anyone ever thought.

CHAPTER FIVE

Search in Networks

● ● ● ● ● ●

STANLEY MILGRAM WAS ACTUALLY A FIGURE OF CONSIDERABLE controversy for much of his professional life. One of the century's great social psychologists, Milgram displayed a genius for designing experiments that plumbed the mysterious interface between the minds of individuals and the social environment in which they typically operate. The results of these experiments were often surprising, but sometimes they were also disturbing and unwelcome. In his most famous study, Milgram brought into his Yale University laboratory members of the local New Haven community, ostensibly to participate in a study of human learning. On arriving, each participant was introduced to the supposed subject of the experiment and was asked to read to him a series of words that he was to repeat. If the subject made a mistake, he was to be punished by receiving an electric shock, administered by the participant. Each successive mistake was to be met with a shock of increasing voltage, rising eventually to harmful and even lethal levels. Meanwhile, the subject would moan, scream, beg for mercy, and thrash about in his restraints. Participants who balked or protested at what they were being required to do to another human being would be instructed to continue by a stern, white-coated, clipboard-toting supervisor. Critically, they were never forced to do anything and were never threatened with reprisal. If at any point they refused to continue, the experiment was terminated without consequence.

Of course, the experiment was all a show. There were no real shocks, and the subject was just an actor. The real point was to see what free-willed individuals would do to another person when they perceived themselves to be obeying orders. The participants were eventually told all this, but during the experiment they thought it was for real, which makes their behavior all the more discomforting. In one variation of the test, during which the participants were instrumental in the experiment, but the shocks themselves were delivered by an intermediary, thirty-seven out of forty participants raised the voltage to lethal levels, leading Milgram to the chilling conclusion that bureaucracies that distanced individuals from the ultimate consequences of their actions were most effective in dispensing brutality. In another variation, the participant was required to hold the subject's hand on an electric plate as he was being shocked! Even today, it is hard to read *Obedience to Authority*, Milgram's elegant account of this work, without pausing for an occasional shudder. But in the postwar ideological landscape of 1950s America, Milgram's findings defied belief, and the experiment became a focus of national outrage.

Although supremely controversial, this experiment did propel Milgram into the pantheon of public intellectuals whose work is so widely remembered and frequently recounted that it has become embedded in the culture itself. We are still shocked (so to speak) by Milgram's experimental results, but we don't question their authenticity, even though his experiments have never been repeated. (In fact, under today's human subjects regulations, they couldn't be.) Nor do we generally question his research on the small-world problem (from chapter 1), even though we continue to find his results intriguing and surprising. Everybody has heard of "six degrees of separation," but most people are unaware of the origin of the phrase, and very few have ever looked closely at Milgram's actual results. Even researchers who cite Milgram's original paper, and who you might think would have scrutinized it carefully, have tended simply to accept his conclusions at face value.

There's a subtle point about science in this behavior. On the one hand, the strength of the scientific project resides in its cumulative nature. Scientists come to their particular problem with an accepted body of knowledge behind them, and on which they expect to draw, without questioning the validity of each and every method, assumption, or set of facts that they use. If we all tried to work everything out from first principles, or even insisted on understanding every piece of the puzzle in equal detail, none of us would ever get anywhere. So to some degree we have to accept that whatever has been acknowledged by the relevant community has been done carefully and correctly, and so can be relied on.

On the other hand, scientists are every bit as human as other professionals and are invariably motivated by many more factors than the unadulterated search for scientific truth. Partly because of their human failings and partly because the truth itself can be so hard to discern, scientists make mistakes, misinterpret their results, or allow them to be misinterpreted by others. Anticipating the inevitability of such errors, the system employs a number of mechanisms, like peer review, academic conferences and seminars, and the publication of dissenting papers, that filter out many of the impurities. But the process is far from perfect, and once in a while we are surprised to discover that a piece of knowledge we had long taken for granted is questionable or even wrong.

SO WHAT DID MILGRAM REALLY SHOW?

THE PSYCHOLOGIST JUDITH KLEINFELD STUMBLED ONTO WHAT now seems a classic instance of such misplaced faith while she was teaching her undergraduate psychology class. She was casting around for a hands-on experiment that her students could perform and that would give them a sense of the applicability of what they were learning in lectures to their lives outside the classroom. Milgram's small-world experiment seemed like a perfect candidate, and Kleinfeld decided she would get her students to redo it in twenty-first-century style, using e-mail instead of paper letters. As it turned out, she never actually got around to it. In preparation for the experiment itself, Kleinfeld started by reading Milgram's papers. Rather than setting a firm base for her experiment, however, Milgram's results—scrutinized carefully—seemed only to raise discomforting questions about his own.

Remember that Milgram started his chains with roughly three hundred people, all of whom were trying to get their letters to a single target in Boston. The story that everyone tells has the three hundred people living in Omaha, but a closer look reveals that one hundred were actually in Boston! Furthermore, of the almost two hundred in Nebraska, only one-half were randomly selected (from a mailing list that Milgram bought). The other half were all blue-chip stock investors, and the target, of course, was a stockbroker. The famous six degrees is an average over these three populations, and as you might expect, the number of degrees varies quite a bit between them, with the Boston natives and the stock investors managing to complete chains more successfully and with fewer links than the random Nebraska sample.

Remember also that the surprising finding about the small-world claim is that anyone can reach anyone—not just people in the same town or people with strong common interests, but anyone anywhere. So really the only population that satisfied, even remotely, the conditions of the hypothesis as it is usually stated (even by Milgram himself) was the ninety-six people picked out of the Nebraska mailing list. At this point, the numbers start to get disturbingly small: of ninety-six starting letters in that population, only eighteen reached the target. *Eighteen!* This is what all the fuss is about? How could anyone have inferred from a mere eighteen chains directed to a single target a principle as universal and all-encompassing as the one we started out trying to explain? And how could all the rest of us have gone along with this, never seriously challenging the plausibility of the claim in the first place?

Perturbed by these questions, Kleinfeld went in search of subsequent papers by Milgram and other authors, assuming that the apparently unsupportable gap between empirical results and their subsequent interpretation had been shored up elsewhere. Again, she was surprised to find that it had not—quite the contrary in fact. Although Milgram and his collaborators did perform other experiments—most significant among which was between a white population in Los Angeles and black targets in New York City—they were limited in much the same way as the original. Even more surprising, only a handful of other researchers had attempted to replicate Milgram's findings, and their results were even less compelling than his. One experiment, for example, sought to connect senders and targets within the same midwestern university—hardly a test of a global principle!

Increasingly bothered by what she was finding, Kleinfeld wound up in the archives of Yale University, digging through Milgram's original notes and unpublished writings, still convinced she must be missing something. She was. As she discovered, Milgram had performed another study in parallel with the Omaha one. This one used starters in Wichita, Kansas, and the wife of a Harvard Divinity School student as the target. Milgram did mention this study in his first paper, published in *Psychology Today*, because it produced the shortest chain he ever measured: the first letter arrived at the target in only four days and used only two intermediaries. What Milgram did not mention in that or any paper is that this first letter was one of only three out of the sixty letters ever to arrive. Kleinfeld also uncovered reports of two follow-up studies in which chain completion rates were so low that no results were published. Kleinfeld's eventual conclusion was that the small-world phenomenon, as it is usually presented to us, has no reliable empirical basis at all.

As this book goes into production, we are currently conducting what is by far the largest ever small-world experiment, in a long overdue attempt to settle the matter. Using e-mail in place of letters, and coordinating the messages through a centralized Web site, we can handle volumes of senders and data of which Milgram could only dream.

At this moment, we have fifty thousand message chains originating in over 150 countries seeking out eighteen targets in the United States, Europe, South America, Asia, and the Pacific. From a college professor in Ithaca (you will never guess who) to an archival inspector in Estonia, from a policeman in Western Australia to a clerk in Omaha, our targets span the gamut of Internet users, a globally dispersed population of half a billion people. Our senders, meanwhile, were recruited via press reports on the experiment that appeared all over the world, and hundreds more contact us every day.

As large as it sounds, however, half a billion people is still not the whole world. And almost certainly, the people who have access to a computer (and enough spare time to use it) represent a relatively narrow cross section of global society. Clearly then, the results even of such a huge experiment won't be *universally* applicable. Furthermore, the experiment is suffering from a problem that Milgram also experienced but not nearly to the same extent—apathy. Today, far more so than in the 1960s, people receive a lot of junk mail, especially e-mail, and are frequently reluctant (or simply too busy) to participate, even when they are asked to by a friend. The result is a crushingly low completion rate—less than 1 percent of all chains that start out ever reach their targets (Milgram, remember, obtained a completion rate of 20 percent). So although we retain high hopes for our experiment, the jury is still out and may well remain so even when our results are fully analyzed. Prehaps the real message then is that the claim of the small-world phenomenon is an incredibly difficult one to resolve empirically.

IS SIX A BIG OR A SMALL NUMBER?

WHERE DOES THIS LEAVE US? AFTER ALL, WE HAVE SPENT A GOOD while trying to understand the small-world phenomenon. Surely we're not going to start questioning it now? Not exactly, but there's an important difference between the small-world phenomenon that we defined

for our network models and the small-world phenomenon as investi-
gated by Milgram—one that we have glossed over until now.
Remember that the main motivation for our original approach to the
problem was the difficulty of empirical verification, so the continued
scarcity of empirical evidence per se doesn't necessarily present a problem for our results. The real issue is that there is a big difference
between two people being connected by a short path (which is all the
small-world network models claim) and their being able to find it.
Recall that Milgram's subjects were supposed to pass the letter along to
one person who they thought was closer to the target than they were.
What they were not supposed to do was send copies of the letter to
every person they knew. Yet that is precisely the kind of computation
that Steve and I conducted in our numerical experiments, and that is
implicit in our statements about shortest path lengths. Therefore, it is
entirely possible for us to be living in a small world, in the sense of the
small-world network models of chapters 3 and 4, yet still doubt the
veracity of Milgram's findings.

Another way to express the difference between our test for a small
world and Milgram's is as a contrast between a *broadcast search* and a
directed search. In broadcast mode, you tell everyone you know, they in
turn tell everyone they know, and so on until the message reaches the
target. According to those rules, if there is even one short path connecting source and target, one of those messages will find it. The downside
is that the network is being completely saturated with messages, as
every single nook and cranny gets probed as a potential path to the destination. This doesn't sound very pleasant, and it isn't. In fact, it's precisely how some of the more troubling computer viruses operate, and
we'll have more to say about them in chapter 6.

Directed searches are a good deal more subtle than broadcast
searches and exhibit different pros and cons. In a directed search like
Milgram's experiment, only one message is being passed at a time, so if
the length of a path between two random individuals is, say, six steps,
then only six people receive the message. If Milgram's subjects had

conducted broadcast searches, sending out messages to every single person they knew, letters would have been received by every person in the entire country—roughly 200 million people at the time—just to reach a single target! Although a broadcast method in principle would have found the shortest path to the target, it would have been impossible in practice. By requiring the participation of only six people, the directed search method avoids overwhelming the system, but the task of finding a short path becomes considerably more involved. Even if in theory you are only six degrees away from anybody else in the world, there are still six billion people in the world and at least that many paths leading to them. Confronted with this maze of mind-boggling complexity, how are we to find the one short path that we are seeking? Well, it's hard—at least on your own.

Long before the Kevin Bacon Game came along, mathematicians used to play a similar game with Paul Erdős. Erdős, being not only a great (and extremely prolific) mathematician but also something of a celebrity in the mathematics community, was thought to be the center of the math world in much the same way that Bacon was for the world of movie actors. As a result, if you have published a paper with Erdős, you get to have an *Erdős number* of one. If you haven't published a paper with Erdős but you have written one with someone who has, then you have an Erdős number of two. And so on. So the question is, "What is your Erdős number?" and the object of the game is to have the smallest number possible.

Of course, if your Erdős number is one, then the problem is trivial. And even if you have an Erdős number of two, it isn't too bad. Erdős was a famous guy, so anyone who had worked with him would probably have mentioned it. But when the Erdős number becomes more than two, the problem gets hard, because even if you know your collaborators rather well, in general you don't know everybody else with whom they have collaborated. If you spent a while on it, and you didn't have too many collaborators, you might be able to write down a reasonably complete list of their other collaborators, if only by looking up all their papers or by asking them. But some scientists have been writing papers for forty years or more and might have accumulated several dozen collaborators in that time, some of whom they might not recall easily. Already this is sounding difficult, but it gets worse—at the next step, you are essentially lost. Imagine trying to write down a list not only of all your collaborators and all their collaborators, but also all of those people's collaborators! You don't even know most of these people, and maybe you haven't even heard of them, so how could you possibly know with whom they have worked? Basically you can't.

What we have tried to do here is effectively a broadcast search in a collaboration network, and again we find that in practice, it is next to impossible. So what everyone ends up doing is a directed search. You pick one of your collaborators whose work you think might be most similar to Erdős's, and then you pick up one of that person's coauthors who you think is closest to Erdős, and so on. The problem is that unless you are an expert in one of the specific fields in which Erdős worked, you might not know which of your collaborators is the best choice. In which case, maybe you guess wrong at the start and wind up stuck in a blind alley. Or maybe you guess right to start with, but one of your subsequent guesses is wrong. Or maybe you are on the right track, but you give up before going far enough. How are you to know how well the search is progressing?

There doesn't appear to be any easy answer to this question, the fundamental difficulty being that you are trying to solve a global problem—finding a short path—using only local information about the network. You know who your collaborators are, and you may even know who some of their collaborators are, but beyond that you are dealing with a world of strangers. As a result, it is impossible to know which of the many paths leading away from you gets to Erdős in the least number of steps. At each degree of separation, you have a new decision to make and no clear way to evaluate your options. Just as someone living in Manhattan might drive east to LaGuardia Airport in order to take a flight to the West Coast, the optimal choice of network path might initially take you in what appears to be the wrong direction. But unlike the drive to the airport, you don't have a complete map of the route in your mind, so the equivalent of driving east to fly west is not so obviously a good idea.

As small as it sounds at first, six can therefore be a big number. In fact, when it comes to directed searches, any number over two is effectively large, as Steve discovered one day when a reporter asked him what his Erdős number was. He figured it out eventually—it's four—but he wasted two full days in the process. (I remember because I was trying to get him to do something else and he was too preoccupied even to talk.) If this sounds like just another way for mathematicians to avoid doing a real job, directed searches do have a serious side. From surfing links on the Web to locating a data file on a peer-to-peer network, or even trying to find the right person to answer a technical or administrative question, we frequently find ourselves searching for information by performing a series of directed queries, often running into frustrating dead ends or wondering if we might have taken a shorter route. As we will see in chapter 9, finding short paths to the right information becomes particularly important in times of crisis or rapid change, when problems need to be solved in a hurry and no one has a clear idea of what is needed or who has it. And as we discovered with the original small-world problem, a simple theory can sometimes tell us a lot about a complex world that we might never have guessed by looking directly at the world itself.

THE SMALL-WORLD SEARCH PROBLEM

THIS TIME, THE KEY BREAKTHROUGH WAS MADE BY A YOUNG computer scientist named Jon Kleinberg, who attended Cornell and MIT, spent a few years working at IBM's Almaden Research Center near San Francisco, and then returned to Cornell as a professor. Kleinberg asked a question that had never occurred to Steve or me, although, as with scale-free networks, it seemed so natural in retrospect that we wondered how we could possibly have missed it. Instead of focusing on the mere existence of short paths, like Steve and I had done, Kleinberg wondered how individuals in a network could actually *find* those paths. The motivation once again was Milgram. Judith Kleinfeld's misgivings aside, clearly some of Milgram's subjects did get their letters to the intended target, and it wasn't obvious to Kleinberg how they had managed to do it. After all, Milgram's senders were essentially trying to perform a directed search in a very large social network about which they had very little information—much less than even a mathematician trying to compute his Erdős number.

The first thing Kleinberg figured out, in fact, was that if the real world worked anything at all like the models that Steve and I had proposed, then directed searches of the kind that Milgram observed should have been impossible. The problem, it turned out, arises from a feature of our small-world models that we haven't discussed yet. Although the models allow us to construct networks that exhibit a variable amount of disorder, the randomness is actually of a special sort. Specifically, whenever a shortcut is created through one of our random rewirings, one neighbor is released and a new neighbor is chosen *uniformly* at random from the entire network. In other words, every node is equally likely to be chosen as the new neighbor, regardless of where it is located or how far away it is.

Uniform randomness seemed like a natural assumption for us to make for our first attempt at the problem, because it doesn't depend on anybody's particular idea of distance. But what Kleinberg pointed out was that people do, in fact, have quite strong notions of distance that they use all the time to differentiate themselves from others. Geographical distance is an obvious one, but profession, class, race, income, education, religion, and personal interests also frequently factor into our assessments of how "distant" we are from other people. We use these notions of distance all the time when identifying ourselves and others, and presumably Milgram's subjects used them too. But because uniform random connections like those in Figure 3.6 don't make use of these notions of distance, the resulting shortcuts are hard for directed searches to utilize. The absence of any reference to the underlying coordinate system—the ring lattice in the case of the beta model in chapter 3—prohibits the search from zeroing in effectively. So the message ends up either jumping around randomly or crawling its way along the lattice. If that had been the case in Milgram's experiment, his chains would have been hundreds of links long, little better than if the message had been passed from door to door all the way from Omaha to Boston.

So what Kleinberg considered was a more general class of network models in which random links are still added to an underlying lattice, but where the probability of a random link connecting two nodes decreases with their distance-apart as measured on the lattice. To keep things simple, he considered the message-passing problem on a two-dimensional lattice (Figure 5.1), on top of which he imagined adding random links according to a probability distribution repre-

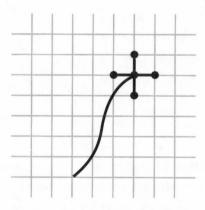

Figure 5.1. Kleinberg's two-dimensional lattice model. Each node is connected to its four nearest neighbors on the lattice and a single random contact.

sented by one of the functions in Figure 5.2. Mathematically, each one of these straight lines plotted on a log-log scale is a power law with an exponent, gamma, that changes from line to line. An exponent of zero (the horizontal line at the top) implies that all nodes in the lattice are equally likely to be random contacts; in other words, Kleinberg's model reduces to a two-dimensional version of the beta model from chapter 3.

So when gamma equals zero, plenty of short paths exist, but as we just saw, they can't be found. When gamma is large, by contrast, the probability of a random shortcut decreases so rapidly with distance that only the nodes that are already close (on the lattice) stand any chance of getting connected. In this limit, each random connection contains a lot of information about the underlying lattice, so paths can be navigated easily. The problem is that because long-range shortcuts are effectively impossible, there are no *short* paths to find. In neither limit, therefore, does the model yield searchable networks. But what Kleinberg wanted to know was, what happens in the middle?

Figure 5.2. The probability of generating a random contact as a function of lattice distance (*r*). When the exponent gamma (γ) is zero, random contacts of all lengths are equally likely. When gamma is large, only nodes that are close on the lattice will be connected.

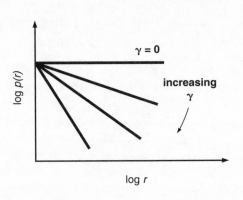

Actually something quite interesting happens. Figure 5.3 shows the typical number of hops required for a message to locate a random target, as a function of the exponent gamma. When gamma is much less than two, the network runs into the same problem as the original small-world models: short paths exist, but they can't be found. And when gamma is much greater than two, short paths simply don't exist. But when gamma equals exactly two, the network achieves a kind of optimal balance between the navigational convenience of the lattice and the distance-cutting power of long-range shortcuts. It is still true that the probability of connecting to any *particular* node will decrease with distance. But it is also true that the greater the distance, the more nodes there are to connect to. What Kleinberg showed is that when gamma is at its critical value two, these conflicting forces exactly cancel each other out. The result is a network with the peculiar property that individuals possess the same number of ties at all *length scales.*

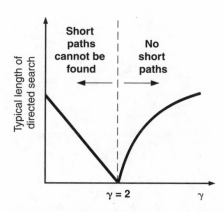

Figure 5.3. Kleinberg's main result. Only when the exponent gamma (γ) equals two does the network possess short paths that individuals can actually find.

This concept is slightly tricky to understand, but Kleinberg came up with a nice image that captured it: Saul Steinberg's "View of the World from 9th Avenue," which graced the cover of *The New Yorker* magazine in 1976 and is reproduced as Figure 5.4. In the picture, 9th Avenue takes up roughly as much space as an entire city block, which in turn occupies the same space as the portion of Manhattan west of 10th Avenue and the Hudson River combined. The same fraction of the picture is then devoted to the entire United States west of the Hudson, to the Pacific Ocean, and then finally to the rest of the world.

Steinberg was making a social comment about New Yorkers' tendency to place as much emphasis on local affairs as on the great matters of the planet—to view themselves as the center of the Universe—but in Kleinberg's model, the image takes on a more tangible meaning. When gamma equals its critical value of two, an individual on 9th Avenue is likely to have the same number of friends in each region, or scale, of the picture. In other words, you should expect to have as many friends who live in your neighborhood as live in the rest of the city, the same number of friends who live in the rest of the state, the same number again in the remainder of the country, and so on, right up to the scale of the whole world. You are roughly as likely to know someone who lives on another continent as someone who lives down the street. Of course, several billion people live on other continents, and probably only a few

Figure 5.4. Saul Steinberg's "View of the World from 9th Avenue," which appeared on the cover of *The New Yorker* in 1976, illustrates Kleinberg's concept of *phases* of search. Private collection, New York.

hundred live down the street. But the idea is that you are so much less likely to know any one *particular* person on the other side of the world that "the rest of the world" and "down the street" end up accounting for more or less the same number of your social acquaintances.

The essence of Kleinberg's result is that when this condition of equal connectivity at all length scales is satisfied, not only does the network exhibit short paths between all pairs of nodes but also individual senders can find the paths if each of them simply forwards the message to the one friend of theirs who seems closest to the target. What makes the search problem feasible is that no one person has to solve it on his or her own. Rather, at each step, all a particular sender has to worry about is getting the message into the next *phase* of its search, where a phase is like the different regions in Steinberg's drawing. So if your ultimate target is a farmer in Tajikistan, you don't have to figure out how to get the message all the way to its destination, or even into the right country. You need only get it to the right part of the world, and then let someone else worry about it. In so doing, you are assuming that the next person in the chain, being closer to the target, has more precise information than you do, and so is better able to advance the search to its next phase. Effectively this is what the condition gamma equal to two guarantees. When the condition is satified by the network, only a few senders are ever required to get a message from one phase into the next—from anywhere in the world to the right country, from anywhere in the country to the right city, and so on. And because the world, just like Steinberg's view of it, can always be partitioned into a small number of these phases, then the total length of the message chain will be short also.

The *Kleinberg condition,* as we came to call it, along with his proof of the impossibility of searching in uniformly random small-world networks, was really a significant step forward in our thinking about networks. Kleinberg's deep insight was that mere shortcuts are not enough for the small-world phenomenon to be of any actual use to locally informed individuals. In order for social connections to be useful—in the sense of finding anything deliberately—they have to encode information about the underlying social structure. But what Kleinberg's model does not explain is how the world could actually be this way. Maybe it's true that if the ties in a social network are arranged in precisely such and such a fashion, then the world suddenly becomes searchable. But how on earth would the network ever get arranged that way in the first place? From a sociological point of view, in fact, the Kleinberg condition seems quite unlikely. Kleinberg, of course, wasn't seeking to make a sociologically realistic model, and by keeping his model as simple as he did, he was able to understand its properties in a way that would have been impossible for a more complicated version. But it did leave the door open for a new way of thinking about the problem, one that incorporated some sociological thinking.

SOCIOLOGY STRIKES BACK

MARK AND I WERE TALKING ABOUT THE DIRECTED SEARCH PROBLEM one day when he was visiting me at Columbia, where I had moved from MIT in August 2000 to join the Sociology Department. After some discussion, we convinced ourselves that the Kleinberg condition wasn't the right way to understand Milgram's results. But how could that be? Hadn't Kleinberg *proved* that any network not connected equally at all length scales couldn't be searched effectively? Well, yes and no. Yes, if it were true that people measure all distances between each other in terms of an underlying lattice. But perhaps what his result was really telling us was that people don't actually compute distances in that fashion. As we strolled around the campus in the spring sunshine, we came up with an example involving the archetypical small-world challenge: how to reach a Chinese peasant farmer. Perhaps neither of us knew any mainland Chinese peasant farmers, and no matter how many of them there were, perhaps we never would. But we knew someone who could at least point us in the right direction.

Erica Jen, a Chinese American, who until recently had been vice president for research at the Santa Fe Institute and who had hired both Mark and me, had attended Beijing University during the years of the Cultural Revolution, long before her arrival at Santa Fe. Furthermore, back in those days, she had been something of a social activist (and one of the first Americans to study at Beijing). We figured that even if she didn't know any rural leaders in the Sichuan province (or wherever our hypothetical farmer lives), she might know someone who did. At any rate, if we were to give a letter to her, we felt pretty confident that in one fell swoop, it would get all the way to China. We didn't know exactly how, and we didn't have any idea what would happen once it got there. But if Kleinberg was right, that wasn't our problem—all we had to do was get the letter to the next phase of its delivery (that is, into the right country) and then let someone else worry about zeroing in on the target.

The difference between Kleinberg's model and our imaginary chain of senders was that although Erica was clearly a crucial link in the chain, and probably the one who would move the letter the farthest distance, she was not, so far as Mark and I were concerned, a "long-range" contact. All three of us had belonged at some point to the same, small, tight-knit community that comprised the live-in researchers of the Santa Fe Institute. From our perspective, it didn't matter where she had lived or what she had been doing twenty years earlier, only that when we knew her, she was our boss, colleague, and friend, working at the same place and interested in many of the same intellectual projects. She wasn't any more distant from us than we were from each other, and for all we knew, her friends in China might in her eyes be no less close to her than we were. In other words, our letter could make what would seem to each of holders to be two small hops—one from us to Erica, and one from her to a friend in China—that, if viewed as a single step, would seem to be very long indeed.

How is it that two short steps can amount to anything other than something short? In a normal lattice model like the ones that Steve and I, and later Kleinberg, had considered, it cannot, which is why all such models (even Kleinberg's) require some fraction of long-range links. Nonetheless it appears that it *can* happen in the real social world, and this paradox has been a persistent source of concern among mathematically inclined sociologists. As long ago as the 1950s, when the mathematician Manfred Kochen and the political scientist Ithiel de Sola Pool first teamed up to think about the small-world problem, social distances appeared to violate a mathematical condition known as the *triangle inequality*, illustrated in Figure 5.5. According to the inequality, the length of any side of a triangle is always less than or equal to the sum of the lengths of the other two sides. In other words, taking one step and then taking another step can never take you any farther from your starting point than two steps. Yet this is precisely what our hypothetical message seemed to have done.

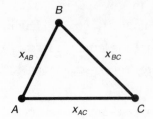

Figure 5.5. The triangle inequality states that the distance $x_{AC} \leq x_{AB} + x_{BC}$. Hence, two short steps can never amount to a long step.

Do social networks really violate the triangle inequality? And if not, why do they appear to? The key to understanding the paradox of distance in social networks is that we can measure "distance" in two different ways, and we tend to get them confused. The first way—the one we have talked about for most of this book—is distance through the network. According to this notion, the distance between two points, A and B, is simply the number of links in the shortest path connecting them. But this is not the definition of distance we typically use when we think about how close or far away we are from somebody else. Instead, as Harrison had reminded me at the AAAS conference in D.C. the previous year, we tend to identify ourselves and others in terms of the groups, institutions, and activities with which we are affiliated.

Having worked on affiliation networks for some time by that stage, Mark and I were familiar with the idea of social identity. But what we realized now was that individuals don't simply belong to groups. They also have a way of arranging them in a kind of social space so as to measure their similarity or differences with others. How they do this is actually somewhat similar to the Steinberg picture shown in Figure 5.4. Starting out at the level of the whole world, individuals break it down, or *partition* it, into a manageable number of smaller, more specific categories. Then they break each of these categories down to a number of subcategories, and each of these into smaller, more specific categories still. This continues on, yielding an image of an affiliation network something like that in Figure 5.6.

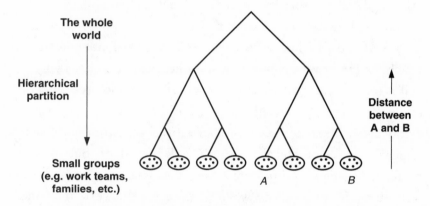

Figure 5.6. Hierarchical partitioning of the world along a single social dimension. The distance between A and B is the height of the lowest common-ancestor group, which in this case is three (individuals in the same bottom-level group are considered to be distance one apart).

The bottom level of this hierarchy comprises the groups that define our closest affiliations—our apartment building, our place of work, or our pastimes. But unlike the affiliation networks of chapter 4, where two actors either belonged to the same group (hence, were affiliated) or not, now we can allow for affiliations of different strengths. Two people may work in different teams but still belong to the same department. Or possibly they are in different departments but belong to the same division, or maybe just the same firm. The higher up the hierarchy one has to go to find a common grouping, the more distant two individuals will be. And, just as in Kleinberg's model, the more distant they are, the less likely they are to know one another. So the equivalent in our model of Kleinberg's gamma exponent was what we called the *homophily parameter,* after the sociological term for the human tendency of like to associate with like. In a highly homophilous network, only individuals who share the smallest of groups can be connected, yielding a fragmented world of isolated cliques. And when homophily is zero, we get the equivalent of the Kleinberg condition, where individuals make associations at all scales of social distance with equal likelihood.

Social distance, therefore, functions in much the same way as it does in Kleinberg's model. But now there are many kinds of distance to which we might refer when assessing the likelihood that two people will meet. Whereas Kleinberg's lattice effectively locates individuals solely in terms of their geographical coordinates, individuals in the real world derive their notions of distance from an assortment of social dimensions. Geographical location is still important, but so is race, profession, religion, education, class, pastimes, and organizational affiliations. In other words, when we partition the world into smaller and more specific groups, we make use of *multiple dimensions simultaneously.* Sometimes geographical proximity is critical, but at other times working in the same industry, going to the same college, or loving the same kind of music may be far more significant in determining who a person knows than where that person lives. Furthermore, being close in one dimension doesn't necessarily imply closeness in another. Just because you grew up in New York doesn't make you any more likely to be a doctor versus a teacher than if you grew up in Australia. Nor does belonging to the same profession necessarily imply that you should live near others in your profession.

Finally, if two people are close in only one dimension, they may consider themselves to be close in an absolute sense even if they are quite distant in other dimensions. You and I need only one thing in common—only a single context for interaction—and that may be enough for us to know each other. Social distance, in other words, emphasizes similarities over differences, and herein lies the resolution to the small-world paradox. As we can see in Figure 5.7, two individuals, A and B, can each perceive themselves as close to a third actor, C, where A is close in one dimension (say, geography) and B is close in the other (say, occupation). Because only the shortest distance counts, it doesn't matter that C is also quite distant from both A and B in some other

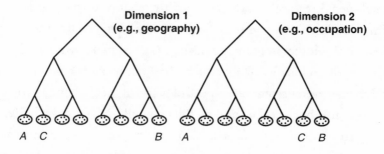

Figure 5.7. Individuals simultaneously partition the world according to multiple, independent social dimensions. This schematic example displays the relative positions of three individuals—A, B, and C—in two dimensions (say, geography and occupation). A and C are close geographically, and B and C are close in occupation. Hence, C perceives itself to be close to both A and B, but A and B perceive each other as distant, violating the triangle inequality of Figure 5.5.

respect. But because A and B are distant in *both* dimensions, they *do* perceive each other as far apart. It's like when you have two friends who you know through different circumstances, and although you like them both, you feel they would have nothing in common. But they *do* have something in common—you—and so whether they know it or not, they are close. Another way to think about this property is that *while groups can be categorized easily, individuals cannot be.* Social identity, therefore, exhibits a multidimensioned nature—individuals spanning different social contexts—that explains the violation of the triangle inequality in social distance. It seemed to Mark and me that the multi-dimensional nature of individual social identity was also what enables messages to be transmitted via a network even in the face of what might appear to be daunting social barriers.

Mark and I got about this far in our discussion before he had to go back to Santa Fe, at which point we both became too busy to do any more work on the problem. About six months later, Jon Kleinberg was visiting Columbia to give a talk in the Sociology Department on his small-world research, and I grabbed the chance to run our ideas past him. Not only did he agree that our approach seemed like the right way to think about the problem, but also he had started down a similar train of thought on his own. This was bad news for us. Jon, after all, is the proverbial rocket scientist—the kind who will hear about a problem in a lecture for the first time and by the end will understand it better than the lecturer. So if he was considering our approach—and according to him, others were too—there wasn't much time to get our act together.

Fortunately, Jon is almost as generous as he is smart, and he agreed to sit on the details of our discussion for a few months to give us a chance to publish something first. Even so, both Mark and I were fully occupied for the foreseeable future, so if we were going to accomplish something in a hurry, we would need some help. Thankfully, also present at the discussion with Jon was Peter Dodds, a mathematician at Columbia who was part of my research group. Peter and I were already working together on another problem (one that we will encounter in chapter 9), so I knew that he could program a computer almost as fast as Mark. And with Mark back in Santa Fe, Peter was a good deal closer! Within days of Kleinberg's visit, Peter and I had dropped our other projects to work on the search problem, and a few weeks later we surprised Mark with a set of results that were even better than we had expected.

Our main finding was that when we allowed the individuals in our model to make use of multiple social dimensions, they were able to find randomly chosen targets in very large networks with relative ease, even when their associations were highly homophilous. In fact, as we can see in Figure 5.8, the existence of searchable networks turns out not to depend too much on the homophily parameter or even the number of social dimensions. In graphical terms, this means that searchable networks exist for any choice of the model's parameters that falls inside the shaded region of Figure 5.8. The equivalent of the Kleinberg condition, by contrast, is the singular point in the extreme bottom left corner of the figure. So our result was, in a sense, the opposite of Kleinberg's. Whereas his condition specifies that the world has to be a very particu-

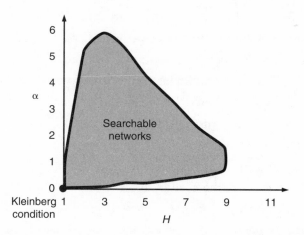

Figure 5.8. Social networks will be searchable whenever they full within the shaded region of the model's parameter space. This region corresponds to social groups being homophilous ($\alpha < 0$), but individuals judging similarity along multiple dimensions (H). The Kleinberg condition, by contrast, holds only at a single point in the bottom left corner of the space of networks.

lar way in order for small-world searches to succeed, our result suggests that it can be almost any way at all. As long as individuals are more likely to know other people like them (homophily), and—crucially—as long as they measure similarity along more than one social dimension, then not only will short paths exist between almost anyone almost anywhere, but also individuals with only local information about the network will be able to find them.

What was doubly surprising, however, was that the best performance was achieved when the number of dimensions was only about two or three. Mathematically, this made some sense. When everyone is using only a single dimension (say, geography) to parse the world, they can't take advantage of their multiple affiliations to hop large distances in social space. Thus, we land right back in Kleinberg's world, where ties have to be arranged equally at all length scales for directed searches to work. And when everyone spreads their contacts out among too many dimensions—when none of your friends belong to the same groups as your other friends—then we are back in the world of random networks, where short paths exist but can't be found. So it made sense that searchable networks should lie somewhere in the middle, where individuals are neither too unidimensional nor too scattered. But that the optimal performance should be when the number of dimensions is about two was still a happy surprise, because that is the number that people actually seem to use.

Several years after Milgram published his seminal small-world paper, another group of researchers led by Russell Bernard (an anthropologist) and Peter Killworth (an oceanographer, of all things) conducted what they called a "reverse small-world experiment." Instead of sending out packets and tracking their progress, as Milgram had done, they simply described the experiment to several hundred subjects and asked them what criteria they would use to direct a packet if they were asked to do so. What they found is that most people use only a couple of dimensions—most dominant were geography and occupation—to direct their message to the next recipient. That this same number should pop out of our analysis twenty-five years later, and with no particular prompting (we had no idea what it would be, but we didn't think it would be two), struck us as pretty remarkable. But we were able to go one better.

By plugging into our model rough estimates of the parameters as they would have applied to Milgram's experiment, we were able to compare our predictions with Milgram's actual results. Figure 5.9 shows the comparison. Not only do the two sets of results *look* roughly comparable, but they cannot be distinguished from each other using standard statistical tests. They are, for all intents and purposes, the same. Given the enormous liberties that our model takes with the complexities of the world, this result was a real stunner. To see how it works, let's return to the example of the hypothetical Chinese peasant farmer. In choosing our friend Erica as our first intermediary, we are making use of two sets of information. First, our notion of social distance leads us to conclude

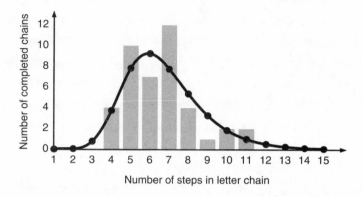

Figure 5.9. The results of the social network search model compared with Milgram's Nebraska results. The bars represent the forty-two completed chains that started in Nebraska, and the curve is the average over many simulated searches preformed according to our model.

that we are quite distant from our target. But it also tells us what groups someone would need to belong to in order to be close. Our notion of social distance thus helps us identify conditions that make an individual a good candidate for passing the message to. And second, we make use of our local knowledge of the network to determine if any of our friends satisfy any of these conditions—that is, do any of our friends belong to at least one group that makes them closer to the target? Erica having lived in China therefore makes her a good candidate.

This is essentially the method that Milgram's subjects used, so what our model shows is that as long as they had at least two dimensions along which they could judge their similarity to others, then even in a world where most of their connections were to people who were much like them, they could still find short paths, even to remote and unfamiliar individuals. That the agreement between our model and Milgram's results should hold so robustly, largely independently of how we pick the particular parameters, tells us something deep about the social world. Unlike networks of power generators or neurons in the brain, individuals in social networks have their own ideas about what makes them who they are. In other words, each individual in a social network comes with a social identity. And by driving *both* the creation of the network *and* the notions of distance that enable individuals to navigate through it, social identity is what leads networks to be searchable.

SEARCH IN PEER-TO-PEER NETWORKS

SEARCHABILITY IS, THEREFORE, A GENERIC PROPERTY OF SOCIAL networks. By breaking the world down the way we do—according to multiple simultaneous notions of social distance—and by breaking the search process itself down into manageable phases, we can solve what seems to be a tremendously difficult problem (just try playing Six Degrees of Kevin Bacon without the computer) with relative ease. Like most insights, the realization that networks have to come from somewhere, and that their origin in social identity is critical to the properties they subsequently display, seems obvious now. But in a science increasingly dominated by physicists, the reentry of sociology into the picture was a significant intellectual development. What we've learned is that while there is nothing wrong with simple models, for any complex reality there are many such models, and only by thinking deeply about the way the world works—only by thinking like sociologists *as well as* like mathematicians—can we pick the right one.

But there is also a practical reason to understand directed searches in networks—namely, the process of finding a target person in a social network, through a chain of intermediate acquaintances, is essentially the same as finding a file or other piece of uniquely specified information in a distributed database. Quite a lot of attention has been paid recently to the potential of so-called *peer-to-peer* networks, particularly in the music industry. The first generation of such networks, the archetype of which is the infamous Napster, was actually only a peer-to-peer network in a limited sense. While the files themselves are located on individuals' personal computers—called *peers*—and the file exchanges occur directly between peers, a complete directory of all available files (and their locations) is maintained on a central server.

In principle, a central directory renders the problem of finding information trivial, even in a very large network—simply query the directory, and it will tell you the location of the file. But central directories are expensive to set up and maintain. From a user's point of view, Internet search engines like Google act like central directories, and generally they do a reasonable job of locating information (the occasional frustration notwithstanding). But Google is not like any ordinary Web site. In order to handle the enormous processing demands of millions of simultaneous queries, it comprises tens of thousands of high-end servers. When I heard Larry Page, one of Google's founders, talk about the company a couple of years ago at a conference in San Francisco, he said they were adding almost thirty servers a day just to keep up with the demand! Central directories may be an effective solution to the search problem, but they're not cheap. Centralized design can also prove quite vulnerable, as the users of Napster discovered when their favorite mechanism for swapping music files was shut down by an irate recording industry. And like an airline network with only one hub through which all flights must pass, once the center goes, the entire system falls apart.

Before Napster had even entered its death throes, however, even more radical forms of distributed databases—what we might call *truly peer-to-peer networks*—had begun to appear in the Internet underworld. One of these, called Gnutella, was designed by a rebellious AOL programmer, who posted the protocol on AOL's Web site sometime in March 2000. Recognizing the potential for copyright infringement implicit in any file-sharing system, and cognizant of their newly consummated merger with Time Warner, AOL authorities removed the offending code within half an hour of its posting. But they were too late—it had already been downloaded and was racing like a drug through the bloodstream of the hacker community, spawning dozens of upgrades and variations. One of Gnutella's early proselytizers was a young software engineer, Gene Khan, who claimed that Gnutella was the answer to every file swapper's prayers and the unstoppable nemesis of the recording industry. Because Gnutella was nothing more than a protocol, it could not be confiscated. And because there was no center, there was no one to sue and nothing to shut down. Listening to Khan, one would have thought that Gnutella was indestructible and all-powerful.

A year later, Khan had been proved half right. No one had succeeded in destroying Gnutella, but then again there didn't seem to be much need. Gnutella had apparently tied itself in knots mainly on account of the same completely distributed architecture that had made it seem so promising. Because no one server knows where all the files are—because there is no central directory—every query becomes a broadcast search that effectively asks each and every node in the network, "Do you have this file?" So a peer-to-peer network like Gnutella, comprising ten thousand nodes, for example, will generate roughly ten thousand times as many messages as a Napster-like network of the same size, where each query is sent only to a single, high-capacity server. Because the aim of peer-to-peer networks is to become as large as possible (in order to increase the number of available files), and because the larger the network is, the worse its performance will be, could it be that truly peer-to-peer networks are inherently self-defeating?

A hint of a Gnutella-like world was revealed by accident a year or so ago by Mrs. Janet Forrest's sixth-grade social studies class at Taylorsville Elementary School in North Carolina. Embarking on an "e-mail project," Mrs. Forrest and her students sent out a sweet little message to all their families and friends, along with the request that all recipients of their message forward it to "everyone you know so that they can send it on to everyone they know (and so on)." They also requested that each recipient respond to them so they could keep a record of how many people had been reached and where. Bad idea. By the time the project was finally cancelled a few weeks later, the class had received over 450,000 responses from every state in the union and eighty-three other countries. And that is just the people who responded! Now imagine that every sixth-grade social studies class in the world tried a similar experiment at the same time. (Unbelievably, I *did* receive another similar message recently from a school in New Zealand, endorsed by none other than the New Zealand minister for education. Some people never learn.) Even worse, imagine that anytime *anyone* wanted to get a message to someone else, they initiated exactly this kind of worldwide broadcast. The age of the Internet would come to a swift and inglorious end, snarled in more traffic than a Bangkok freeway.

In general, therefore, central directories are expensive and vulnerable, and broadcast searches generate more problems than blessings. As a result, efficient search algorithms that require only local network information should be of considerable practical interest. So it is one of the most intriguing aspects of the small-world phenomenon that individuals embedded in social networks appear able to solve the peer-to-peer search problem, even if they themselves don't know how they do it. By understanding and exploiting the properties of the sociological version of the problem, we therefore can hope to devise novel solutions for network search problems that don't necessarily involve people at all. Complementing our approach to the problem, some other solutions to the directed search problem on peer-to-peer networks have been proposed to take advantage of other aspects of network structure. Most notable among these efforts is that of the physicist Bernardo Huberman and his student Lada Adamic at the HP research lab in Palo Alto, California.

Observing that the degree distribution of the Gnutella network appears to follow a power law (over some range), Adamic and Huberman proposed a search algorithm whereby nodes direct queries to their *most highly connected* neighbor, which then checks its local directory, and the directories of its neighbors, for a copy of the requested file, repeating the procedure if the file cannot be found. In this manner, each query quickly locates one of the relatively small number of hubs that are characteristic of a scale-free network and that together are connected to most of the network. By searching this *hub network* randomly, the group demonstrated that most files could be found in a relatively short time without overburdening the network as a whole. As ingenious as this approach is, it suffers from a weaker version of the central directory solution—hubs must have much higher capacity than ordinary nodes, and the performance of the network depends sensitively on the operability of the critical hubs. By contrast, the searchability of social networks appears to be a highly egalitarian exercise. In our model, ordinary individuals are capable of finding short paths, so no special hubs are required.

Prehaps the main point, however, is that in stimulating novel solutions to (apparently) such different problems, the small-world problem provides the perfect example of how the different disciplines can help each other build the new science of networks. Back in the 1950s, Kochen (a mathematician) and Pool (a political scientist) were the first to think about it but couldn't find a solution without computers. Milgram (a psychologist), aided by White (a physicist-sociologist) and followed by Bernard (an anthropologist) and Killworth (an oceanographer), then attacked the problem empirically but couldn't explain how it actually worked. Thirty years later, Steve and I (mathematicians) turned the problem into one about networks generally but failed to see its algorithmic component, leaving that door for Jon (a computer scientist) to open. Jon, in turn, left the door open for Mark (a physicist), Peter (a mathematician), and me (now a sociologist of sorts) to walk through and pick up the solution that now seems to have been lying there all along.

It's been a long trail, almost fifty years, and now that we think we finally understand the problem, it seems like someone ought to have figured it out long ago. But *it had to happen this way.* Without Jon, for example, we would never have understood how to think about the search problem—we would not have known which door to walk through. And without our early work on small-world networks, Jon would never have thought about the problem in the first place. Without Milgram, none of us would have known what it was that we were trying to explain. And without Pool and Kochen, Milgram would have been off doing a different experiment. In retrospect, everything seems obvious, but the truth of the matter is that the small-world problem could only have been solved through the combined efforts of many different thinkers, coming at it from all different angles and bringing with them an incredible diversity of skills, ideas, and perspectives. In science, just as in life, one cannot simply fast-forward the tape to see what the ending looks like, because the ending is written only in the process of getting there. And like a successful Hollywood movie, the end, even when it brings a certain sense of resolution, is merely a prologue to the sequel. For us, the sequel was dynamics. And next to the mysteries of dynamics on a network—whether it be epidemics of disease, cascading failures in power systems, or the outbreak of revolutions—the problems of networks that we have encountered up to now are just pebbles on the seashore.

CHAPTER SIX

Epidemics and Failures

THE HOT ZONE

Most of us, it's probably fair to say, don't lose a lot of sleep over the possibility of catastrophic epidemics. That's probably because most of us have not read *The Hot Zone,* Richard Preston's true story of Ebola, an astonishingly lethal virus that kills its victims in a blood-gushing finale of such heartless fury that only nature could have devised it. Named after the Ebola River, which drains the northern districts of what used to be Zaire but is now the Democratic Republic of Congo, the virus first emerged from its hiding place in the jungles in 1976. It struck first in Sudan and then two months later in Zaire, where it broke out in fifty-five villages almost simultaneously, claiming nearly seven hundred lives that year alone.

Although surprisingly little is known about it, Ebola is thought to have jumped, like HIV, from monkeys to humans and comes in at least three strains of increasing deadliness. A recent outbreak of Ebola in Uganda was of the Sudan strain, which, with a kill rate of a mere 50 percent, is the ninety-pound weakling of the family (Ebola Zaire claims 90 percent of its victims). Even so, 173 people died in the district of Gulu between October 2000 and January 2001 before the outbreak ran its course. Other outbreaks over the past three decades have killed similar numbers in roughly similar circumstances, mostly small, isolated villages with poor health care facilities. The horror stories of these outbreaks are legion: victims coming to the local doctor, complaining of flulike symptoms; crashing and bleeding several days later in the nearest medical clinic; the terrible realization, usually too late, that Ebola has struck; heroic medical workers cut down in the first line of defense; widespread panic; dozens of bleeding bodies discovered in deserted huts; villages left devastated and abandoned; whole regions terrorized. Ebola is truly a monster, a messenger sent straight from Hell.

Ironically, Ebola's tremendous violence is also its one weakness: it is literally too deadly for its own good. Unlike silent, insidious HIV, Ebola exhibits all the subtlety of a train wreck, revealing its true nature in a matter of days and killing shortly thereafter. Furthermore, once the symptoms appear, the victims are so incapacitated and so obviously ill that they have difficulty traveling and can be quarantined with relative ease, thus reducing the virus's ability to spread to new hosts. As a result, the majority of outbreaks have been contained in remote areas near the rain forest and away from major population centers.

Only once, during the second outbreak in 1976, did Ebola make its way to the big city, when a young nurse known only as Mayinga N., infected with the Zaire strain, spent the day wandering around Kinshasa, the capital and largest city of Congo. Fortunately catastrophe was averted by another quirk of the virus: Ebola, at least in its early stages, is not all that contagious. Even when a patient is in a terminal state, hemorrhaging internally and coughing blood-infused mucus into the air, it is generally thought that the virus can only reach new hosts through a break in their skin, or through a permeable membrane like those in the nose or eyes. By the time nurse Mayinga had reached that stage, however, she had already realized her fate and was quarantined in the hospital.

Reading this, you might think that Ebola is just another line item in the seemingly endless litany of horrors afflicting sub-Saharan Africa. And Africa, that most exotic and tragic of continents, is surely distant enough that the next plague, if and when it comes, need not affect us any more dramatically than an occasional wince as we flip through the morning newspaper. If *The Hot Zone* has one thing to teach you, however, it is that you can stop relaxing right now. Ebola is a problem not just for Africa but for the whole world. Just as HIV crawled its grisly way down the Kinshasa highway from its birthplace in the jungles and somehow, probably in one of the coastal cities, found Gaetan Dugas—the Canadian flight attendant, better known as *patient zero*—who thereby brought it to the bath houses of San Francisco and introduced AIDS to the Western world, so too could the right chain of events free Ebola from its shackles.

More so even than Preston's vivid descriptions of Ebola-induced death, it is the potential for a global explosion of the virus that is the most disturbing aspect of his account. Over the last century, not only have we humans intruded deeply into the ancient ecologies of the African rain forests, where the most deadly viruses lie in wait, but also we have built an international system of transportation networks that can transmit an infectious disease to the world's metropolises and power centers within a few days—less time, it so happens, than Ebola's incubation period. Preston even says of one of his doomed characters, as he sits on a small plane to Nairobi vomiting bagloads of black blood, "Charles Monet and the life form inside him had entered the net."

The prospect of Ebola showing up in the local shopping mall is almost too horrible to contemplate, but after reading *The Hot Zone*, you're almost amazed that it hasn't. In fact, the main subplot of the book describes the outbreak of a third strain of Ebola among a population of monkeys at an Army research lab in Reston, Virginia, just outside Washington, D.C. The virus, now identified as Ebola Reston, turned out to be harmless to humans but spectacularly lethal to the poor monkeys, all of whom died. But Ebola Reston is so similar to Ebola Zaire that none of the standard tests at the time could tell them apart, and for a few nail-biting days, Zaire is what the scientists and animal handlers who had been exposed to it thought it was. Had it actually been the Zaire strain—and it is pure dumb luck that it wasn't—then we would all know a lot more about Ebola today than we do.

VIRUSES IN THE INTERNET

THESE DAYS BIOLOGICAL VIRUSES ARE NOT THE ONLY POTENTIAL source of epidemics, as Claire Swire discovered much to her chagrin right before Christmas 2000. Claire Swire is a young Englishwoman who a few days earlier had apparently had a brief affair with a young Englishman named Bradley Chait. Being a modern woman, she sent him an e-mail the following day, complimenting him in a way that Chait found so appealing, he decided to share it with his friends. Only his best friends, mind you—only six of them. But they apparently found the compliment so entertaining that each of them forwarded it to several of *their* nearest and dearest, many of whom, it turns out, felt the same way. And so it went, this little e-mail, with Chait's one-line amendment, "That's a nice compliment from a lass," around and around the world, amusing roughly 7 million readers in a matter of days. *Seven million!* Poor Claire had to go into hiding to avoid the ensuing press frenzy, and Chait was "disciplined" by his law firm for unauthorized use of his e-mail account (as if people don't send personal e-mail from work all the time). It's a silly story maybe, but a fine example of the power of exponential growth, especially when mixed with the near costless transfer of information afforded by the Internet. And on this topic, there are plenty of serious things to say.

Viruses, both human and computer, essentially perform a version of what we have been calling a *broadcast search* throughout a network. Broadcast searches, as discussed in chapter 5, represent the most efficient way of starting from any given node and finding every other one by systematically branching out from each newly connected node to each of its unexplored neighbors. When a disease embarks on a "search," however, it isn't looking for anything in particular—it is simply seeking to spread itself as far and wide as possible. So "efficiency" for an infectious entity like a virus generally carries with it connotations of mayhem. The more contagious a virus is, and the longer it can keep the host in an infectious state, the more efficient it is at searching. Ebola, therefore, is more efficient than HIV in that it is significantly more infectious (HIV-infected patients don't vomit blood in the emergency room), but is less efficient in that it kills so quickly. And both HIV and Ebola are far less efficient than the influenza virus, which not only keeps its hosts alive for much longer but also is able to spread via airborne particles. To put the importance of disease efficiency in perspective, if Ebola *were* contagious via airborne transport, modern civilization might well have come to an end sometime in the late 1970s.

As much as we should be concerned about the possibility of a human "slate wiper," as Preston dubs truly devastating plagues, in terms of efficiency alone computer viruses are far more troublesome than human ones. A virus—whether human or computer—can be regarded as little more than a set of instructions for reproducing itself, using material from the host as its building blocks. In humans the immune system screens out foreign and possibly dangerous sets of instructions, but computers don't generally possess immune systems. In essence, the function of a computer is to execute instructions as efficiently as possible, regardless of where the instructions came from. So they are considerably more vulnerable to malicious bits of code than are people. And although a worldwide computer epidemic might not signal the end of civilization, it could still exert a significant economic toll. No such event has occurred yet, but we have already experienced some disquieting tremors. In the last few years of the twentieth century, even before Y2K turned out to be the biggest anticlimax of the millennium, a series of computer virus outbreaks caused a significant level of disruption and inconvenience to hundreds of thousands of users around the world. Government agencies, large corporations, and even the usually ambivalent public began to sit up and take notice.

Computer viruses have been with us for decades, so why have we only recently started to experience them on a global scale? The answer, as it is for so many questions about the second half of the 1990s, is the Internet. Before the Internet, viruses circulated and computer users experienced occasional difficulties. But back then, pretty much the only way to transmit a virus from one machine to another was via a floppy disk, which had to be physically inserted into the machine. Certainly it was possible for the contaminated disk to circulate among many computers, and once a computer was infected, saving related files onto an uninfected floppy would infect that disk as well. So the potential for exponential growth clearly existed, but the largely manual nature of spread—like Ebola's requirement for a break in the skin— generally reduced the efficiency of the virus enough that small out- breaks did not become full-fledged epidemics.

The Internet, particularly e-mail, has changed all that, as the world began to understand in March 1999 with the arrival of the Melissa virus. Although Melissa was generally referred to as a virus (or a bug), it actu- ally had a lot in common with another type of malicious code known as a *worm*. Worms wreak havoc not so much on individual computers as on networks of computers. They replicate and transmit themselves in large numbers from machine to machine without being activated by a user. Melissa, which at the time was the fastest-spreading virus that anyone had ever seen, arrived in the form of an e-mail with the subject line, "Important message from <name>," where <name> was that of the user sending the message. The body of the message said, "Here is the docu- ment you asked for . . . don't show anyone else ;-)" and a Microsoft Word document called list.doc was attached. If the attachment was opened, Melissa's macro would automatically mail copies of itself to the first fifty addresses in the user's e-mail address book. If any of the addresses hap- pened to be a mailing list, everyone on the list would get the virus.

The results were quite dramatic. First detected on Friday, March 26, Melissa had spread all over the globe within hours, and by Monday morning it had infected over one hundred thousand computers in three hundred organizations, bombarding some sites with so many messages (in one case, thirty-two thousand messages in forty-five minutes!) that they were forced to take their mail systems off-line. It could have been a lot worse, however. Melissa not only was relatively benign—its worst action was to insert a harmless reference to *The Simpsons* into an open document if the minute of the hour matched the day of the month—but also was only able to propagate via the Microsoft Outlook mail program. Users without Outlook could still receive the virus but were unable to pass it on, a distinction that has important consequences for the likelihood of a truly devastating global virus (and possibly even for the Microsoft corporation itself) as we will discuss later. First, however, we have to learn a thing or two about the mathematics of infectious disease. In particular, we need to understand better the conditions under which a small outbreak of a disease becomes an epidemic.

THE MATHEMATICS OF EPIDEMICS

MODERN MATHEMATICAL EPIDEMIOLOGY WAS BORN OVER SEVENTY years ago with the introduction of the *SIR model,* which was formulated by two mathematicians, William Kermack and A. G. McKendrick, and is still the basic framework around which most infectious disease models are constructed. The letters in the acronym represent the three primary states (illustrated in Figure 6.1) that any member of a population can occupy with respect to a disease: *susceptible,* meaning that the individual is vulnerable to infection but has not yet been infected; *infectious,* implying that the individual not only is infected but also can infect others; and *removed,* implying that the individual either has recovered or has otherwise ceased to pose any further threat (possibly by dying). New infections can only occur when an infected individual, often called an *infective,* comes into direct contact with a *susceptible.* At that point, the susceptible can become infected, with a probability that depends on the infectiousness of the disease and the characteristics of the susceptible (some people, clearly, are more susceptible than others).

Obviously, who comes into contact with whom will depend on the network of associations in the population. To complete the model, therefore, we must make some assumptions about that network. The standard version of the model, for example, assumes that interactions between members of the three subpopulations occur

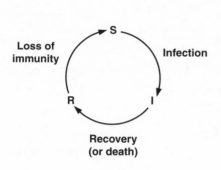

Figure 6.1. The three states of the SIR model. Each member of the population can be susceptible, infected, or removed. Susceptible individuals can become infected by interacting with infectives. Infectives can either recover or die, thus ceasing to take part in the dynamics. If they recover, they might become susceptible again through loss of immunity.

purely at random, as if all the members of the population were being stirred in a large vat, like the one in Figure 6.2. As the image of the vat suggests, pure randomness isn't a very good proxy for human interactions, but it certainly simplifies the analysis considerably. In the SIR model, the randomness assumption implies that the probability of an infective meeting a susceptible is determined solely by the size of the infected and susceptible populations—in a vat, there is no population structure to speak of. The problem still isn't trivial, but now at least it is possible to write down a set of equations whose solutions depend only on the size of the initial outbreak and a few parameters of the disease itself, like its infectiousness and the recovery rate.

According to the model, when an epidemic does occur, it should follow a predictable course known to mathematicians as *logistic growth.* As Figure 6.3 indicates schematically, each infection requires the participation of both an infected and a susceptible individual. Hence, the

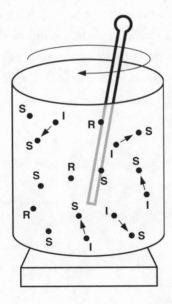

Figure 6.2. In the classical version of the SIR model, interactions are assumed to be purely random. One way to think of random interactions is as individuals being mixed together in a large vat. The main consequence of the random mixing assumption is that interaction probabilities depend only on the relative population sizes, a feature that greatly simplifies analysis.

rate at which new infections can be generated depends on the size of both populations. When the disease is in its early stages, the infected population is small, and therefore so is the rate of new infections—as the top diagram in Figure 6.3 shows, there just aren't enough infectives to cause much damage. This *slow-growth phase* is also the most effective stage in which to prevent an epidemic, as even a few averted infections can drive the disease back into remission. Unfortunately, an epidemic in its early stage can be hard to distinguish from a random grouping of unrelated cases, especially if public health authorities are poorly coordinated or reluctant to admit that they have a problem.

By the time the density of infectives becomes too great to be overlooked or ignored, the epidemic has typically entered the *explosive phase* of logistic growth (middle diagram in Figure 6.3). Now there are many infected and many susceptible individuals, so the rate at which new infections occur is maximized. Epidemics in the midst of explosive growth are essentially impossible to stop, as British farmers witnessed in 2001 when foot-and-mouth disease raged for half a year throughout most of England and parts of Scotland. When the epidemic was detected in mid-February, only three weeks after the first cases had occurred, forty-three farms had already been affected. That might seem like a lot of farms, but the epidemic was still in the initial,

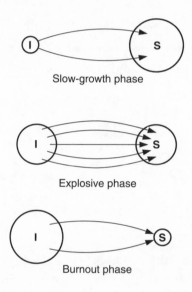

Slow-growth phase

Explosive phase

Burnout phase

Figure 6.3. In logistic growth, the rate of new infections depends on the size of the susceptible and infected populations. When either population is small (top and bottom diagrams), new infections are rare. But when both populations are intermediate in size (middle diagram), infection rates are maximized.

slow-growth phase. By September, the number of farms suspected of infection had grown to over nine thousand, despite the preventative slaughter of nearly 4 million sheep and cattle.

Eventually, however, even the most out-of-control epidemics come to an end, if for no other reason than they burn themselves out. Because there are only so many people (or in the case of foot-and-mouth disease, animals) who can be infected, susceptible targets become harder and harder to find, and the trajectory of the disease flattens off again. This is the *burnout phase* of logistic growth. In the foot-and-mouth epidemic, this self-limiting process was accentuated by the effective quarantine of farmlands and the massive extermination of animals (only about two thousand actual cases of the disease were ever detected, a tiny percentage of the number killed). From start to finish, therefore, the course of an epidemic displays a characteristically S-shaped curve, like the one in Figure 6.4. That the main features of this trajectory— slow growth, explosion, and burnout—are explicable in terms of the logistic growth model suggests that the forces governing an epidemic, when it occurs, are fundamentally quite simple.

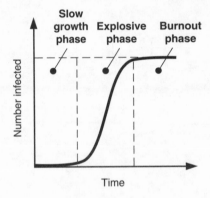

Figure 6.4. Logistic growth, displaying the slow-growth phase, explosive phase, and burnout phase.

But epidemics do not always occur. In fact, most outbreaks of disease either are contained by human intervention or (much more often) burn themselves out before infecting more than a tiny fraction of the population. As terrifying as it was, the Ebola outbreak of 2000, for example, doesn't qualify as a genuine epidemic. Although 173 victims is a significant number in absolute terms, the outbreak remained confined to a geographically localized collection of villages, never seriously threatening the bulk of the potentially vulnerable population. The foot-and-mouth epidemic of 2001, by contrast, affected almost the entire country. In terms of the SIR model, stopping an epidemic is roughly equivalent to preventing it from reaching the explosive growth phase of Figure 6.4, which in turn implies focusing not on the size of the initial outbreak but on its *rate of growth*. The key measurement of a disease in this respect is its *reproduction rate,* the average number of new infectives generated by each currently infected individual.

The mathematical condition for an epidemic is that the reproduction rate of the disease has to be greater than one. If the reproduction rate is kept below one, then infectives are removed from the population faster than new ones are generated, and the disease will die out without becoming an epidemic. But if the reproduction rate exceeds one, then the disease increases not only its spread but also the speed with which it will continue to spread, and explosive growth inevitably commences. The knife edge between these two conditions, where a single host passes on its burden to precisely one single new host, is called the *threshold* of an epidemic. Preventing an epidemic amounts to keeping the reproduction rate below its threshold.

In the classic SIR model, where population structure is ignored, the reproduction rate, and therefore the epidemic threshold, is determined entirely by the properties of the disease itself (its infectiousness, and the speed with which infectives recover or die) and by the size of the susceptible population with which its hosts can interact. Thus, safe-sex practices have constrained the HIV epidemic in some parts of the world by targeting its infection rate, while the widespread extermination of animals in Britain during the foot-and-mouth epidemic very likely reduced its severity by limiting the effective size of its susceptible population.

That the threshold reproduction rate in the classical model should be exactly one turns out to be one of those deep convergences that makes mathematics so interesting. The epidemic threshold is, in fact, exactly analogous to the critical point at which a giant component appears in a random network (see chapter 2), where the reproduction rate is mathematically identical to the average number of network neighbors. And the size of the infected population as a function of the reproduction rate (Figure 6.5) is exactly analogous with the size of the giant component in Figure 2.2. The onset of an epidemic, in other words, occurs when the disease passes through exactly the same phase transition that Erdős and Rényi discovered in their apparently unrelated problem about communication networks. This remarkable similarity, however, also suggests an obvious criticism. If we rejected random graph models as realistic representations of real-world networks, social or otherwise, should we not also reject any conclusions

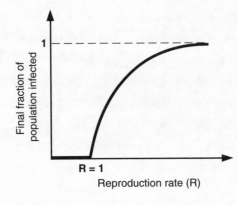

Figure 6.5. Phase transition in the SIR model. When the *reproduction rate* (R) of the disease exceeds one (the epidemic threshold), an epidemic occurs.

about epidemics that are based on the same assumptions? The dependence of the reproduction rate on the size of the susceptible population alone, for example, doesn't account for any of the features of social or network structure that might be useful in combating an epidemic. As we will see, some lessons of the classical model hold up even in the complex world of networks, but new, network-oriented lessons have to be learned as well.

EPIDEMICS IN A SMALL WORLD

STEVE AND I, REMEMBER, WERE INTERESTED IN DYNAMICS FROM the start. After all, we had gotten ourselves into the network business originally because we were interested in the dynamics of coupled oscillators—the crickets. So once we had some network models to play with, we naturally wondered how different dynamical systems might behave on them. The first such system that we tried to understand was the Kuramoto oscillator model, from chapter 1, on which Steve had done so much work earlier in his career. Unfortunately, as simple as the Kuramoto model is, its behavior on a small-world network was still too complicated for us to make sense of it (a statement that remains true even several years later). So we started looking for a simpler kind of dynamics, and once again, Steve's biological interests came in handy. "The SIR model is about the simplest kind of nonlinear dynamics I can think of," he said one day in his office, "and I'm pretty sure that no one has really thought about the SIR model located on a network—at least not a network anything like this. Why don't we try that?"

So we did, but this time I did some homework first. Sure enough, while the basic SIR model had been generalized in many ways to include the idiosyncrasies of particular diseases and the varying susceptibilities of different demographic groups, nothing like small-world networks had come up in the literature. That much was encouraging, as was the deep equivalence between the classic SIR model and the connectivity of a random graph. Whatever the behavior of a disease in a general small-world network, we could be sure that it had to resemble the classical SIR behavior in the limit where all links had been randomly rewired (as in the right panel of Figure 3.6). So not only did we have a network model, which by this stage we understood reasonably well, we also had a well-established benchmark against which to compare our results.

The first natural comparison to make against the random limit was for a disease spreading on a one-dimensional lattice—the ordered end of the small-world spectrum, and the left panel of Figure 3.6. In a lattice, as discussed in chapter 3, the links between nodes are highly clustered, implying that a spreading disease is continually being forced by the network back into the already infected population. As displayed in Figure 6.6, in a one-dimensional lattice, a growing cluster of infectives actually consists of two kinds of nodes—those in the interior of the cluster (who cannot infect any susceptibles) and those on the boundary, or *disease front*. No matter how large the infected population is, the size of the disease front remains fixed; hence, the *per capita* rate of growth of the infected population inevitably decreases as the infection spreads.

Thus, a lattice presents a very different context for an epidemic than does the random mixing assumption above. It also makes the reproduction rate difficult to compute, so we decided to compare the results for our different networks directly in terms of infectiousness. And the difference was striking. As shown in Figure 6.7, the same disease,

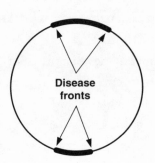

Figure 6.6. On a ring lattice, the disease front (where infectives and susceptibles interact) is fixed. As the size of the infected population increases, more infectives lie in its interior, where they cannot reach new susceptibles. Hence, diseases spread slowly on lattices.

spreading in a lattice, tends to infect far fewer people than it does in a random graph, and there is no longer any clear threshold. The take-home message is that when diseases are confined to spread in only a limited number of dimensions—even, say, by the two-dimensional geography of the land—only the most infectious diseases will develop into true epidemics. And even then, they will be slow, creeping epidemics, rather than explosions, giving public health authorities time to respond and a well-defined area on which to focus.

An example of just such a creeping epidemic is the black death that swept across Europe in the fourteenth century, wiping out about one-fourth of the entire population. As mind boggling a statistic as that is, an epidemic like the plague probably couldn't happen today—at least not in the industrialized world. As the map in Figure 6.8 shows, the plague started in a single town in southern Italy (where it is thought to have arrived on an infected ship from China) and then spread like a ripple along the surface of a pond after a stone is dropped. Because the disease was transported mostly by rats infested with plague-carrying fleas it took three years, from 1347 to 1350, for the front to propagate across Europe. Neither medical science nor public health services at that time were able to prevent the plague's relentless progress, so its relatively slow speed didn't make much difference ultimately. But in the modern world, any disease forced to travel by such slow and inefficient means could be identified and contained.

Unfortunately, diseases today have much better mechanisms for transport than scurrying rats. And no sooner did we permit even a small fraction of random links into our network models, than the relative stability of the lattice model splintered apart. To see this effect,

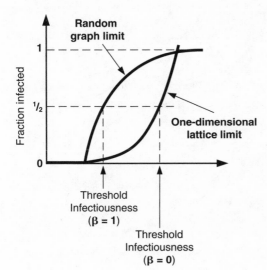

Figure 6.7. The fraction infected versus infectiousness for the random graph limit ($\beta = 1$) and lattice limit ($\beta = 0$) of the beta model from chapter 3. The value for the *threshold infectiousness* represents the infectiousness required for one-half of the population to become infected.

consider the horizontal line drawn halfway up across Figure 6.7. The points at which the two infection curves intersect the line represent the values of the disease infectiousness at which that fraction of the population is infected (in the figure, the fraction is one-half, but we could have chosen some other value). Call this value the *threshold infectiousness* (remember we can no longer use the reproduction rate to define the threshold for an epidemic, so we use a fixed fraction of the population instead), and then ask how it varies with the fraction of random shortcuts in the network. As we see in Figure 6.9, the threshold infectiousness starts high—the disease must be highly infectious in order to contaminate a large population—but then drops rapidly. More importantly, it approaches the worst-case scenario of a completely random network while the network itself is still far from random.

This observation might help explain why epidemics such as the

Figure 6.8. Map of the progression of the plague (black death) in Europe between 1347 and 1350.

foot-and-mouth disease epidemic in Britain can explode so rapidly. Because foot-and-mouth disease spreads between animals either through direct contact or indirectly via wind-blown droplets excreted from symptomatic animals and in virus-laden soil, one might expect any initial outbreak to spread only along the two-dimensional geography of the English countryside, much like the plague did seven hundred years earlier. However, the combination of modern transportation, modern livestock markets (in which animals from geographically dispersed forms are exchanged, or simply come into physical contact), and recreational hikers carrying infected soil on their boots has broken the constraints of geography. As a consequence, British sheep and cattle farms are linked by a network of transportation systems that can move infected animals (and people) anywhere in the nation overnight. And because these links are, for all intents and purposes, random, the virus only needed to find a few of them in order to launch itself into fresh territory. An important early problem in combating the epidemic, for example, was that the forty-three farms on which foot-and-mouth disease was first detected were not neighboring farms. Hence, the virus had to be fought on many fronts simultaneously, with more being added every day.

That the results of the random mixing model turn out to be so easily replicated even in highly clustered networks is not good news for the

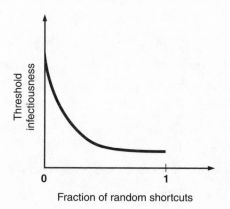

Figure 6.9. The threshold infectiousness required for an epidemic to occur decreases dramatically for small amounts of randomness in a network.

world. If diseases really do spread on small-world networks, then it would seem that we are continually facing a worst-case scenario. Even more troubling, because very few people ever have more than local information about their networks, it can be very hard for public health authorities to make individuals understand the immediacy of an apparently distant threat and thus change their behavior. AIDS is a good example of this problem. For more than a decade after the AIDS epidemic was first identified, HIV infection was generally considered to be confined to only a few, quite specific communities—gay men, prostitutes, and intravenous drug users. So if person X didn't have sex with anyone in these three categories, and none of his or her sexual partners did either, then person X was safe, right? Wrong! What is obvious to us now that we have seen the virus infect almost entire nations in southern Africa is that in the small world of sexual networks, even an apparently distant danger must be taken seriously. Particularly troubling is the thought that HIV was able to breach its initial boundaries at least in part because of the perception that it couldn't.

The phrase "think globally, act locally" is therefore nowhere more appropriate than to the prevention of epidemics. Remember that infectious diseases, unlike the search problems of the previous chapter, conduct what we have called broadcast searches. So if there is a short path through a network of contacts between an infective and a susceptible, it doesn't matter if either person knows it's there or even whether they could find it if they wanted to. Unless the disease is stopped somehow, it will find the path because it is blindly probing the network for *every* path. And unlike Gnutella users or Mrs. Forrest's sixth-grade class from the previous chapter, it is quite happy to overload the entire network with copies of itself—that's what infectious diseases do. That our perception of risk from an infectious disease, whether it be HIV, Ebola, or even, say, the West Nile virus, should be so out of kilter with the actuality of disease transmission is definitely a cause for concern.

The situation is not all gloom and doom, however. As stated earlier, outbreaks of disease typically do not become epidemics, and in this respect small-world networks have something encouraging to teach us. On a small-world network, the key to explosive growth of a disease is the shortcuts. Diseases don't spread very effectively on lattices, and although small-world networks exhibit some important features of random graphs, they still share with lattices the property that locally, most contacts are highly clustered. So *locally,* the growth of a disease behaves very much like it does on a lattice: infected individuals interact mostly with other already infected individuals, preventing the disease from spreading rapidly into the susceptible population. Only when the disease cluster reaches a shortcut—whether that be an Ebola victim getting on a plane, or a truckload of cattle infected with foot-and-mouth driving up the M1—does it start to display the worst-case, random mixing behavior. So unlike epidemics on a random graph, epidemics in a small-world network have to survive first through a slow-growth phase, during which they are most vulnerable. And the lower the density of shortcuts, the longer this slow-growth phase will last.

A network-oriented strategy for preventing epidemics, therefore, would not only attempt to reduce infection rates in an overall sense, it would focus particularly on likely sources of shortcuts. Interestingly enough, the needle exchange program that has been effective in reducing the spread of HIV among intravenous drug users exhibits both these features. Removing dirty needles from circulation eliminates one mechanism by which HIV can spread, thus reducing the overall rate of infection. But it also works by virtue of the *particular* infections that it prevents. Dirty needles are shared not only among friends but also by complete strangers, who might pick up and reuse a discarded needle. In other words, reused needles are a source of random connections in the disease network. Just as the ban on animal movements and the closure of country footpaths throughout England in 2001 reduced the potential for long-range shortcuts, eliminating needles from the system closes off one avenue of escape from the slow-growth phase of an epidemic, giving health authorities a better chance of catching up with the disease.

Thinking about the structure of networks may also explain other subtleties in the spread of a disease that would not be apparent in the absence of a network-oriented approach. Recently the Spanish physicist Romualdo Pastor-Satorras and the Italian physicist Alessandro Vespignani pointed out one such feature of real-world *computer* viruses that classical SIR models have trouble explaining. After studying prevalence data available at a popular on-line virus bulletin, they concluded that most viruses exhibit a peculiar combination of long-term and low-level persistence "in the wild." The combination is peculiar because according to the standard SIR model, every virus must either generate an epidemic (in which case some significant fraction of the population will be infected) or quickly burn itself out. In other words, either it explodes or it doesn't. But unless it happens to have a reproduction rate of exactly one, the critical point of the phase transition in Figure 6.5, it can't just drift around failing to do either. By contrast, many of the 814 viruses whose timelines were recorded on the virus bulletin appeared to do precisely that. Some of them had been floating around for years, despite the availability of antivirus software usually within days or weeks after their initial detection.

Pastor-Satorras and Vespignani proposed an explanation that explicitly included features of the e-mail network through which they hypothesized the viruses were spreading. Taking Barabási and Albert's scale-free model as a proxy for the structure of e-mail networks—an assumption that was supported (albeit inconclusively) a year later in a report by a German team of physicists—the two physicists showed that when spreading on scale-free networks, viruses don't display the same threshold behavior that they do in the standard model. Instead, as in Figure 6.10, the fraction of the population infected tends to grow continuously from zero as the infectiousness of the disease increases. In a

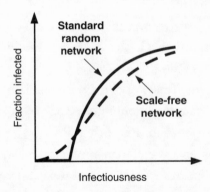

Figure 6.10. Comparison of infection curves in standard versus scale-free random networks. The scale-free networks display no critical point at which epidemics suddenly appear.

scale-free e-mail network, most nodes have only a few links, which is to say that most people send e-mail to only a few others on a regular basis. But a small fraction of e-mail users have very extensive address books, containing a thousand names or more, and are apparently diligent enough to keep up with them all! It is this minority that Pastor-Satorras and Vespignani hypothesized is more or less responsible for the long-term persistence of viruses—only one of them needs to become infected with a virus once in a while for it to continue to circulate at measurable levels throughout the population as a whole.

Apparently then, even the simplest features of real-world networks, like local clustering and scale-free degree distributions, can have important consequences for the spread of disease and, most important, the conditions governing epidemics. The study of disease models is, therefore, an important subfield in the new science of networks. In a world where several tens of millions of people are now infected with HIV and where the prevalence varies, even within Africa, from less than 2 percent to more than one third of a country's population, the importance of understanding the spread of infectious disease in networks cannot be overstated. Much work remains to be done, but already some promising directions are appearing in the network literature. And while the SIR model remains central to this effort, the physicists, true to form, have started tackling the problem in their own way. In particular, they have introduced to the study of epidemics, a set of techniques that goes under the general label of *percolation theory*.

PERCOLATION MODELS OF DISEASE

HISTORICALLY, PERCOLATION THEORY DATES BACK TO WORLD War II, when Paul Flory and his collaborator Walter Stockmayer used it to describe the *gelation* of polymers. If you have ever boiled an egg, then you are familiar with some aspects of polymer gelation. As the egg is heated, the polymers in the egg white link up and bond to each other, one pair at a time. Then at some critical point, the white undergoes a sudden, apparently spontaneous transition, called *gelation*, in which a very large number of branching polymers suddenly become bound together in a single, coherent cluster that spans the entire egg. In breakfast terms, before gelation, the egg is liquid; after gelation, it is solid. The first success of percolation theory was Flory and Stockmayer's explanation of how this transition could happen almost instantaneously, not slowly and incrementally as one might have expected. Although it was originally developed to answer questions in organic chemistry, percolation theory has subsequently proved a useful way of thinking about all sorts of problems, from the sizes of forest fires, to yields from underground oilfields, to the electrical conductivity of composite materials. More recently, it has also been used to think about the spread of disease.

In late 1998, not long after I had arrived at the Santa Fe Institute, I had started talking with Mark about the disease-spreading work that I had done with Steve the year before. Based on a simple SIR model, Steve and I had been able to make some conclusions about the depen-dence of the epidemic threshold on the density of random shortcuts. But we hadn't been able to understand exactly how the mechanism worked, or how the effect of random shortcuts varied with the density of the network. Since then I had been teaching myself some of the basics of percolation theory, which seemed a natural way to go about asking many of the same questions. And Mark, being an expert in statistical physics, was the obvious person to ask. As I soon learned was typical with Mark, once he got interested in the problem, it didn't take very long for results to follow.

Imagine a very large population of individuals (*sites* in percolation terminology) connected to one another by a network of ties (*bonds*) along which a disease might be transmitted. Each site in the network is susceptible or not, with some probability, called the *occupation probability*, and each bond is either *open* or *closed* with a probability that is equivalent to the infectiousness of the disease. The result looks something like the diagrams in Figure 6.11 (although for much larger networks), where the disease can be thought of as an imaginary fluid pumped out of a source site. Starting at the source, the disease will always "flow" along any open bonds that it encounters, spreading from one susceptible site to another until no more open bonds can be accessed to new susceptible sites. The group of sites that can reached in this fashion from a randomly selected start point is called a *cluster*, where the entry of a disease into a given cluster necessarily implies that all the sites in that cluster become infected also.

In the left diagram of Figure 6.11, the occupation probability is high and many bonds are open, implying a highly infectious disease to which most of the population is susceptible. In this condition, the largest cluster spans almost the entire network, thus implying that an outbreak at a random location in the network; would be expected to

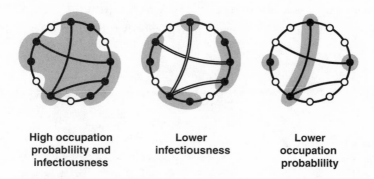

High occupation probablility and infectiousness   **Lower infectiousness**   **Lower occupation probablility**

Figure 6.11. Percolation on a network. Solid circles (links) correspond to occupied (open) sites (bonds). Connected clusters are shaded.

spread widely. In the other two diagrams, by contrast, either the infectiousness (middle diagram) or the occupation probability (right diagram) is low, implying that disease outbreaks will be small and localized, no matter where they occur. In between these extremes lies a complicated continuum of possibilities in which clusters of all sizes can exist simultaneously, and the extent to which a disease spreads is determined by the size of the particular cluster in which it originates. The main objectives of percolation theory are to characterize this distribution of cluster sizes and to determine how it depends on the various parameters in the problem.

In the language of physicists, the possibility of an epidemic depends on the existence of what is called a *percolating cluster*—a single cluster of susceptible sites (connected by open bonds) that permeates the entire population. In the absence of a percolating cluster, we would still see outbreaks, but they would be small and localized. However, a disease that starts somewhere on a percolating cluster, instead of dying out, will spread throughout even a very large network. The point at which a percolation cluster appears, usually referred to as *percolation*, turns out to be exactly analogous to Flory and Stockmayer's gelation in polymers. It is also equivalent to the epidemic threshold in SIR models at which the reproduction rate of the disease first exceeds one (and therefore, by association, the connectivity transition of a random graph). As Figure 6.12 shows, below the threshold, the size of the

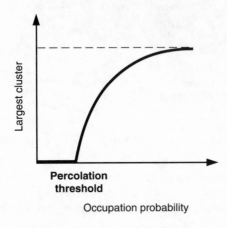

Figure 6.12. Largest infectable cluster in a network. Above the percolation threshold, the largest cluster occupies a finite fraction of the network, implying that an outbreak can become an epidemic.

largest cluster, when viewed as a fraction of the whole population, is negligible. But as the critical point is reached, we observe the sudden and dramatic appearance of a percolating cluster—apparently out of nowhere—through which the disease can spread uninhibited.

The distance through a network that a disease will typically spread before burning itself out is equivalent to what physicists call the *correlation length*, a term we encountered in chapter 2 in the context of global coordination. There, the divergence of the correlation length implied the system had entered a critical state in which even local perturbations could propagate globally. Much the same result is true in percolation models of disease spreading. Right at the percolation transition, the correlation length becomes effectively infinite, implying that even very distant nodes can infect one another. What Mark and I figured out, in the case of small-world networks, was how the correlation length depended on the fraction of random shortcuts. In agreement with the crude results that Steve and I had gotten almost two years earlier, Mark and I showed that even a small fraction of random shortcuts could alter the correlation length dramatically. But now, by solving for the conditions under which the correlation length diverged, we could determine the position of the percolation transition—and thus the epidemic threshold—precisely.

NETWORKS, VIRUSES, AND MICROSOFT

THIS RESULT WAS A PROMISING START AND DEMONSTRATED THAT for some problems at least, epidemics can be better understood by using a percolation approach than with the standard SIR model. Unfortunately, percolation on realistic networks is a difficult (and unsolved) problem, and further progress proved hard to come by. To keep the analysis manageable, for example, most percolation models either assume that all sites in the network are susceptible, and focus on the bonds (this is called *bond percolation*), or assume that all bonds are open, and focus on the sites (*site percolation*). Roughly the same methods work for both kinds of percolation, and in many respects, they behave in a similar fashion. Mark and I, for instance, studied the site percolation version, but shortly thereafter, Mark and another Santa Fe physicist, Cris Moore, extended the results to bond percolation. In some respects, however, site and bond percolation differ significantly, occasionally yielding quite different predictions for the likelihood of an epidemic.

Before racing ahead with the analysis, therefore, one has to think carefully about which version—bond percolation or site percolation—best captures the nature of the disease in question. In the case of a virus like Ebola, for example, it seems reasonable to assume that all people are susceptible, and to focus on the extent to which they can infect each other. Therefore, the relevant formulation of an Ebola-related percolation problem would be bond percolation. Computer viruses like the Melissa bug, however, will generally pass between any susceptible computer and any other computer (all bonds are effectively open), but not all computers are susceptible. So a percolation model of a computer virus probably ought to be of the site percolation variety. Taking the Melissa bug as an example, only a certain fraction of computers in the world are susceptible to the virus because it can only spread via the Microsoft Outlook e-mail program, and not everybody uses Outlook.

Unfortunately for Microsoft users, so many computers run Outlook that the largest connected cluster of them almost certainly percolates. If it didn't, in fact, we wouldn't see global viral outbreaks like Melissa and its protégés, the Love Letter and Anna Kournikova viruses. Universal software compatibility clearly confers some significant benefits on individual users. But from the perspective of system vulnerability, when everybody has the same software, everybody also has the same weaknesses. And every piece of software has weaknesses, especially large, complex operating systems like Microsoft's. In a way, the only amazing thing about Melissa-style outbreaks is that they haven't happened more often. And if they do start to happen more often—if Microsoft software acquires the reputation for persistent vulnerability—then large corporations, and even individuals who cannot afford to have their computers put out of action every time a new virus appears anywhere in the world, may start to look for alternatives.

What should Microsoft do? The obvious approach is to make its products as resilient as possible to attack by any wormlike virus, and in the event of an outbreak, make antivirus software available as quickly as possible. These measures have the effect of reducing the occupation probability of the network, thus shrinking and possibly even eliminating the percolating cluster altogether. But if massive corporations like Microsoft, who are the natural targets of any hacker desiring fame and glory, want to protect their customers and their market share, they may also have to think a little more radically. One solution might be to switch from a single integrated product line to several different products that are developed separately and that are designed *not* to be entirely compatible.

From a conventional software point of view, one that emphasizes compatibility and economies of scale, deliberately disintegrating a product line might seem like a crazy idea. But in the long term (and the long term may not be that long), a proliferation of nonidentical products would reduce the number of computers susceptible to any particular virus, rendering the system as a whole dramatically less vulnerable to the largest of viral outbreaks. This is not to say that Microsoft products would not still be vulnerable to virus attacks, but at least they would not be drastically more vulnerable than the competition. Ironically, a disintegrated product line is more or less the fate that Microsoft appears to have avoided recently in its antitrust battle with the Justice Department. One day Microsoft may be seen as its own worst enemy.

That subtle differences in the mechanism of disease spread can be translated into distinct versions of the general percolation framework—possibly with quite different outcomes—suggests a certain amount of care is required in applying the methods of physics to the problem of epidemics. In the next chapter, in fact, we will see that other distinctions must be made if we are to understand the difference between biological contagion and social contagion problems like the diffusion of a technological innovation, distinctions that again carry important implications for the real-world phenomena we would like to understand. Percolation models, however, are so naturally applied to networks that they will continue to play an important role in the study of network epidemics. And as Mark and I soon realized, percolation is interesting for other reasons as well. Once again, however, it was László Barabási and Réka Albert who were one step ahead.

FAILURES AND ROBUSTNESS

LIKE MOST FEATURES OF COMPLEX SYSTEMS, GLOBAL CONNECTIVITY is neither unambiguously a good nor a bad thing. In the context of infectious diseases or computer viruses, the existence of a percolating cluster in a network implies a potential epidemic. But in the context of a communication network like the Internet, where we would like to guarantee that packets of data will reach their destination in some reasonable time, then a percolating cluster would seem an absolute necessity. From the perspective of protecting infrastructure, therefore, from the Internet to airline networks, it is the *robustness* of the network's connectivity, with respect to accidental failures or deliberate attacks, that we want to preserve. And from this point of view also, percolation models can be extremely useful.

Having shown that a number of real networks like the Internet and the World Wide Web were what they called *scale-free,* Albert and Barabási started to wonder whether scale-free networks bore any competitive advantages over the more traditional varieties. Remember that in a scale-free network, the distribution of degree is governed by a power law instead of the sharply peaked Poisson distribution that we find in uniform random graphs—a distinction that, in practice, translates to a small fraction of "rich" nodes having very many links, and many other "poor" nodes having hardly any at all. Now Albert and Barabási became interested in the question of how well connected two networks, one a uniform random network and the other a scale-free network, would remain once their individual nodes started to fail.

Thinking about network robustness as a connectivity issue mapped the problem neatly into one about site percolation. In this application, however, the occupation probability played the opposite of its role in disease spreading. Whereas Mark and I had been primarily interested in the effect of occupied (susceptible) sites, Albert and Barabási were concerned with unoccupied sites—in network terms, the nodes that had failed. And in terms of robustness, the less effect that each unoccupied site had on the connectivity of the network, the better. Albert and Barabási also had a different view of connectivity from the one that Mark and I had used. Whereas we were concerned only with whether a percolating cluster existed or not, they wanted to know precisely how many steps would be required for a message to cross from one side of the cluster to the other. Neither definition is universally the right way to think about robustness, but theirs was clearly relevant to systems like the Internet, where an increase in the typical number of hops taken by a message increases both its expected delivery time and its likelihood of being dropped.

The first thing that Albert and Barabási showed was that scale-free networks are far more resistant to *random* failures than are ordinary random networks. The reason is simply that the properties of scale-free networks tend to be dominated by the small fraction of highly connected *hub* nodes. Because they are so rare, these hubs are much less likely to fail by random chance than their less connected and far more plentiful counterparts. And like the absence of a minor rural airport from the airline network of the United States, the loss of a "poor" node goes largely unnoticed outside its immediate vicinity. In ordinary random networks, by contrast, the most-connected nodes are not nearly so critical, nor are the less well-connected nodes so inconsequential. As a result, every lost node will be missed—maybe not a great deal but more so than in a scale-free network. Invoking recent evidence that the Internet is in fact scale-free, Albert and Barabási went on to propose their model as an explanation of how the Internet works so reliably, even though individual routers fail all the time.

There is another side to robustness, however, that they also pointed out. Although in some networks like the Internet, router failures do occur at random, failures also can be a consequence of deliberate attacks, which may not be random at all. Even in the Internet, denial-of-service attacks, for example, tend to target highly connected nodes. And in other examples, from airline networks to communication networks, it is the hubs that are clearly the prime targets of any potential saboteur. What Albert and Barabási showed was that when the most highly connected nodes in a network are the first to fail, scale-free networks are actually much *less* robust than uniform networks. Ironically, the vulnerability of scale-free networks to attack is due to exactly the same property as their apparent robustness: in a scale-free network, the most connected nodes are so much more critical to overall network functionality than their counterparts in a uniform network. The overall message, therefore, is an ambiguous one: the robustness of a network is highly dependent on the specific nature of the failures, with random and targeted failures offering diametrically opposite conclusions.

Although both kinds of failure are important to consider, the preferential failure of hubs seems particularly significant because it needn't be either deliberate or malicious. In many infrastructure networks that depend disproportionately on a small fraction of highly connected nodes, higher-than-average failure rates for those nodes may actually be an unavoidable consequence of their connectivity. For example, in the airline network, the massive amount of traffic passing through the major hubs increases their tendency to fail, a phenomenon with which air travelers in New York are painfully familiar. At LaGuardia Airport in Queens, both incoming and outgoing flights are stacked so close together that even a series of trivial delays, which at a small airport would be absorbed by the normal interval between flights, can accumulate to keep planes on the ground for hours, even on a picture-perfect day. In the year 2000, in fact, LaGuardia was the origin for 127 of the 129 most delayed flights in the country! And delays at hubs like LaGuardia are not just a problem for local travelers. Each flight delayed at a major hub tends to generate knock-on delays at its destination airport as well. So the more flights a hub handles, the greater its own chance of experiencing delays, and the greater the chance those delays will reverberate throughout the system.

The heavy dependence of modern airline networks on a subnetwork of hubs, therefore, causes them to be particularly susceptible to occasional widespread delays. But it also suggests a solution. Rather than persisting with a system in which the hubs bear all the burden of getting people from point A to point B, airlines could shift some of the links from the largest, most failure-prone hubs to the smaller regional airports whose delays derive principally from problems originating at the hubs. Under such an arrangement, airports in Albuquerque and Syracuse, for example, would be connected directly, rather than having to route flights through Chicago or St. Louis. Very small airports, like those in Ithaca and Santa Fe, meanwhile would remain spokes. By reducing the effective degree of the hubs, the network as a whole would retain much of the efficiency that it derives from their large scale but would reduce the probability of individual failure. And even in the event that a hub did fail, fewer flights would be affected, causing the system as a whole to suffer less.

As straightforward as it seems in retrospect, Albert and Barabási's result was pretty neat, and with their paper on "Network attack and failure" gracing the cover of *Nature,* it generated a minor storm of media attention. We once again kicked ourselves for missing an obvious problem, and then, with the help of another of Steve's students—Duncan Callaway—scrambled to catch up. Duncan, in fact, succeeded in solving a much more difficult problem than the one Barabási's group had tackled. Using the techniques that Mark, Steve, and I had developed for studying the connectivity of random networks, Duncan managed to compute the different percolation transitions exactly, rather than just using computer simulations. He also managed to solve the problem for both link and node failures, and showed how to apply the model not just to scale-free networks but to random networks with *any* kind of degree distribution at all. All in all, it was an impressive effort, and the four of us managed to get a very nice paper out of it. But ultimately it didn't make a lot of difference. To a rough approximation, our findings were much the same as Albert and Barabási's, and we had to admit, they had thought of it first.

Fortunately for us, applying percolation techniques to the problems of the real world is a somewhat subtle business, so there were plenty of interesting problems left. Not only are real networks more complicated than any random model—scale-free or otherwise—but also the nature of the process itself is often poorly represented by the standard assumptions of percolation theory. Percolation models, for example, typically assume that all nodes have the same likelihood of being susceptible. In reality, however, heterogeneity is an important feature of human and many nonhuman populations. Even in matters like disease spreading, individuals can vary widely in their inherent susceptibility or their capacity to infect others. And when behavioral and environmental factors are considered, the differences across a population can be complicated by the presence of strong correlations. It is often the case in sexually transmitted diseases, for example, that high-risk individuals are significantly more likely to interact with other high-risk individuals, a behavioral characteristic that may have social origins but clearly has epidemiological consequences.

Furthermore, the states of individuals can be correlated not only according to their intrinsic characteristics but also dynamically. A good analogy is the cascading failure in the power transmission grid discussed in chapter 1. If you were to assign failure probabilities to nodes at random, even taking account of their individual differences, you would still be missing an essential part of the problem: the role of *contingency*. The massive failure that occurred on August 10, 1996, remember, was not a result of multiple independent failures but rather was a *cascade* of failures, each one of which made subsequent failures more likely. Cascades of contingent, interdependent failures are more complicated to model than the percolation problems we have dealt with so far, but they happen all the time, and not just in engineering systems like the power grid. In fact, possibly the most widespread and interesting group of cascade problems lies in the realm of social and economic decision making. It is to these important, fascinating, and deeply mysterious problems that we now turn.

CHAPTER SEVEN

Decisions, Delusions, and the Madness of Crowds

• • • • • •

NOT LONG AFTER I HAD LEFT THE PERFECT WEATHER OF SANTA FE
for the rain and sleet of Cambridge, Massachusetts (I arrived smack in
the middle of Hurricane Floyd), I started wondering if the lessons that
Mark and I had learned from our studies on disease spreading could be
applied to the problem of contagion in financial markets. It was the fall
of 1999, and the dot-com bubble was reaching its frenzied peak.
Venture capital money seemed to be flowing freely to anyone with a
remotely plausible business plan, and throughout my temporary home
at MIT's Sloan School of Management, the students could hardly wait
to get out the door and make their fortunes. Start-up fever had risen so
high that Merrill Lynch, traditionally one of the biggest employers of
MIT graduates, was threatening to cancel its annual recruiting drive
because no one was showing up for their presentations!

In the midst of all this excitement, Andrew Lo, a finance economist and my adviser at the time, suggested that I take a look at a book
by Charles Mackay, called *Extraordinary Popular Delusions and the
Madness of Crowds*. As its alluring title suggests, Mackay's book is a
treatise on the many manifestations of mania, from witch trials to cru-

sades, in which usually sensible, often educated people end up behaving in ways that later seem hard to fathom. And mania, as Mackay makes patently clear, has no better friend than money. A year later, after the dot-com crash, you might well have concluded that all those (now unemployed) MBAs, not to mention a good many Wall Street analysts, had indeed been possessed of "extraordinary delusions."

You might, however, have suspected that such widespread hallucinations of illusionary value—not just the late 1990s infatuation with technology, but the Texas savings and loan crisis of the 1980s, the crash of October 1987, the Mexican peso crises, and the bubble economies of Japan and then later Korea, Thailand, and Indonesia—are a relatively recent feature of an increasingly rugged and treacherous financial landscape. Surely in the days before automatic trading systems, round-the-clock markets, and frictionless international capital flows—before even telephones, telegraphs, or transcontinental railways—the rapid proliferation of unfounded belief, and the ready capital to back it, would have been impossible, at least on a large scale. Not so—*Extraordinary Popular Delusions* was published in 1841, and by that year Mackay's subject was already two centuries old.

TULIP ECONOMICS

FINANCIAL CRISES, AS I SOON LEARNED FROM ANDY, ARE AT LEAST as old as the Roman Empire. But the first example of modern times, and one of the titillating stories that Mackay relates, is known as the Dutch Tulip Bubble. In 1634, the year that it began, tulips had only recently been introduced to western Europe from their native Turkey, and apparently came with a good deal of social cachet. The flowers' desirability only being enhanced by their generally difficult and fragile nature, tulip bulbs were already fetching hefty prices in the Amsterdam flower market. But before long, professional speculators and "stock jobbers" got involved, artificially driving up the prices of the bulbs with the intention of reselling them at a later date for even higher prices still. Entranced by the promise of instant wealth and reassured by the massive influx of capital from foreign investors, even ordinary citizens were drawn into the madness, to such an extent that the routine business of the economy was virtually abandoned. According to Mackay, at the height of the boom, a single bulb of the rare species Viceroy was exchanged for "two lasts of wheat, four lasts of rye, four fat oxen, eight fat swine, twelve fat sheep, two hogsheads of wine, four tons of beer, two tons of butter, one thousand pounds of cheese, a complete bed, a suit of clothes, and a silver drinking cup." And if that sounds preposterous, the most prized species of all, the Semper Augustus, could fetch over twice as much. I am not making this up.

With so much real money invested in a commodity of so little objective value, it should come as no surprise that the Dutch began to behave in a way that can only be described as odd, in some cases selling off their entire livelihoods in order to own a root or two of some prized flower. Tangible assets, being relatively worthless, were easy to purchase, and so buying and selling, borrowing and spending all grew rampant. For a short while, as Mackay puts it, Holland became "the very antechamber of Plutus." Of course, it couldn't last. At the end of the day, a tulip is a tulip is a tulip, and even the most deranged Dutchman cannot pretend otherwise forever. The inevitable crash came in 1636. Tulip prices fell to less than 10 percent of their giddy heights only months before, generating even more insanity among the now angry masses as they searched in vain for scapegoats and sought to relieve their rapidly accumulating debt.

Several decades later, but still more than a century before Mackay's book, two other great imperial nations, France and England, were struck almost simultaneously by financial bubbles that resembled the tulip fiasco not only in their origins and basic trajectory but also in the level of absurdity that they inspired in their respective citizenries. This time the object of speculation was stock in two companies, the South Sea Company in England and the Mississippi Company in France, that promised extremely high rates of return by virtue of their access to new and largely untapped frontiers (the South Pacific for the English company and the southern colonies of what was to become the United States of America for the French one). Just as with the tulips, investors flocked and the share prices soared, prompting further speculation, further demand, and further upward pressure on the prices. As with Holland, both England and France became awash with paper money for which more and more real assets were exchanged, leading to universal delusions of great wealth and perpetuating the ever more unstable upward spiral of prices. And just as for the poor Dutchmen, the dot-com investors of the late 1990s, and a good many fools and their money in between, the bubble eventually burst, shattering the illusions and fortunes of a once euphoric population.

FEAR, GREED, AND RATIONALITY

So WHY HAVEN'T WE LEARNED? AFTER THE BETTER PART OF FOUR hundred years, what is it about financial bubbles that always makes the next one seem like it won't burst until it's too late? The cynical answer is that greed and fear are as universal and timeless as any human characteristics, and that once they are aroused, neither thoughtful analysis nor even past experience can compete. Without the promise of vast personal wealth, successful lawyers would not have rushed to work for some start-up company selling groceries over the Internet, any more than a sensible Dutchman would have traded in his Amsterdam apartment for a tulip bulb. And without the panic that only a deep-seated suspicion of ultimate worthlessness can bring, the sudden and almost simultaneous conflagration of so many Internet-related companies (some of which might even have been judged worthy of survival in a more circumspect environment) would not have happened. Like most cynical explanations, however, this one is not very helpful, effectively condemning us to live out the financial equivalent of *Groundhog Day* but without the

Hollywood ending (eventually Bill Murray learns). The cynic claims that we can't change people, and this may be true. But the claim tells us nothing about the mechanics of how financial crises actually work, how one differs from another, or how we might design *institutions* that help people at least live in peace with their demons.

Standard economic theories of decision making, it turns out, are even less helpful. Economics, it must be remembered, is the antithesis of cynicism. People are selfish, it claims, but they are also *rational.* So greed is always mitigated by knowledge, and fear is absent entirely. The result, as Adam Smith famously predicted, is that rational agents, optimizing their selfish interests, will "be led by an invisible hand" to a collective outcome that is at least as good as any other. Governance—which is to say not just *the* government, but institutions, regulations, and externally imposed restrictions of all kinds—is likely only to upset the proper functioning of the market. Although Smith was writing specifically about the political economy of international trade, his logic has subsequently been applied to markets of all kinds, including financial markets, in which, naturally, there will not be any crises.

The basis of this sunny outlook is that according to the standard view of financial traders as rational, optimizing agents, bubbles can't happen without destabilizing speculators to drive them, and destabilizing speculators aren't supposed to exist. Why not? Because destabilizing speculators buy and sell assets not according to any measure of their "true" value but by following trends in prices, typically buying on the upswing and selling on the downswing. As a result, this particular kind of speculator is often called a *trend follower,* as opposed to a *value investor* who only buys an asset when she perceives it to be underpriced (and sells when it is overpriced). So if an asset price rises from its true value for some reason, the trend followers will rush in and buy it, thereby paying more for the asset than it is actually worth. Of course, in doing so, they push the price up further, and by selling the asset at this even more inflated price, they hope to make a profit. For every trend follower who sells at a profit, however, there has to be another one buying it (because no value investor would be interested), and therefore making an even bigger mistake than his predecessor.

Eventually the chain of fools must come to an end, at which point prices will fall and some of the trend followers will lose money. If the price falls far enough, such that it sinks once more below its true value, the value investors will step in and start buying again, thereby reaping a profit at the expense of the trend followers. No matter how many trend followers make profits, those profits can only come at the expense of other trend followers, so overall the population of trend followers will *always* lose money to the value investors. Because the net transfer of wealth from trend followers to value investors is a fundamental property of speculation, no rational person would choose to be a trend follower; hence, market prices should always reflect the true values of the corresponding assets. In the language of finance, markets should always be *efficient*. But what if real people are simply stupid? The theory has an answer to this objection also: even if they are fools, the mere fact that they have an overall tendency to lose money will eventually force them out of the market, by a kind of Darwinian selection. Value investors will take money from the trend followers until the trend followers go broke and leave. In the long run, only the value investors will be left, and order will triumph. No speculation, no overtrading, no bubble.

As logical as it sounds, the inevitable triumph of rationality actually reveals a paradox in the operation of financial markets. On the one hand, fully rational investors in a properly functioning market, by making use of all available information, should converge on a price for every asset that correctly reflects its true value. No one should ever make decisions based purely on movements of the price itself, and if they try to, they will eventually be forced out of the market. On the other hand, if everyone behaves rationally, prices will always track values, so no one, not even value investors, can make a profit. The result is that not only will there be no bubbles, there won't be any trading at all! This is a somewhat problematic conclusion for a theory of markets, as without trade, markets have no way to adjust prices to their "correct" values in the first place.

Another way to view the logic of rationality, therefore—and certainly the perspective suggested by history—is that it is largely irrelevant to whatever is going on in real financial markets. Yes, people try to maximize profits, and yes, speculators often lose money. Possibly it's true that over long enough periods of time, all speculators, even those who occasionally make staggering profits, eventually lose it all. Like gamblers in a casino, some people may win for while, but ultimately the only real winner is the house (which, by the way, might explain why they keep building them). Yet people keep speculating just as people keep gambling.

But if humans are not entirely rational in the strict economist's sense of the word, neither are they completely at the mercy of uncontrollable passions. Even the most outrageous speculators have methods to their madness. And as for the rest of us—most of the time, we are simply trying to get by, making the best of the situations that present themselves and avoiding trouble as much as possible. It doesn't sound like a terribly volatile mixture, and frankly, most of the time it isn't. In fact, while bubbles and crashes tend to get all the attention, the behavior of financial markets is often remarkably calm, even in the face of external events, like changes in government and terrorist attacks, to which one might expect it *would* overreact. So the real mystery of financial markets is not that they are rational or irrational, but that they are both. Or neither. Either way, when large numbers of ordinary people get together, it seems that *most* of the time they behave quite reasonably, but *once in a while* they end up behaving like madmen. And financial crises are but one example of the normally sensible but occasionally extraordinary behavior that groups, mobs, crowds, and even whole societies are capable of displaying.

COLLECTIVE DECISIONS

NOT LONG BEFORE I STARTED LEARNING ABOUT THE ORIGINS OF financial crises, I had been reading up on another subject that fascinated me: the evolution of cooperative behavior. Cooperation is a quality of human behavior so ubiquitous that it is sometimes (mistakenly) thought to be one of the principal differentiators of humanity from mere beasts. But the origin of unforced cooperative behavior is actually a profound paradox that has occupied generations of thinkers from philosophers to biologists. The paradox, in a nutshell, is why self-interested humans should ever behave unselfishly in a world where doing the right thing by others is inherently costly and easily exploited.

Imagine going out to dinner at a nice restaurant with a large group of friends under the assumption that at the end of the night, you will all split the bill. The menu contains a wide range of choices, from an inexpensive but plain pasta dish to an extravagant filet mignon. If everyone orders a fancy meal, it will be an expensive night all around, so naturally you'd be doing everyone a favor by ordering the pasta. On the other hand, if you get the steak and your friends make do with pasta, you will have a great meal at something close to half the price. More to the point perhaps, if you *don't* order the steak and your friends do, you may well end up paying a great deal for a lousy bowl of spaghetti. The question, of course, as you all sit there contemplating your choices, is how much do you care about your own selfish pleasure versus the welfare of your friends?

As quaint as this game might appear at first, the *diner's dilemma*—as it was coined by Natalie Glance and Bernardo Huberman, the two physicists who proposed it—is actually a canonical example of a *social dilemma*. Also known as *public goods games,* social dilemmas deal with a collective good, like recycling services or a mass transit system, the existence of which requires a sufficiently large fraction of the population to contribute to the good despite the existence of easier, more profitable, or more selfish alternatives (for example, driving instead of taking the bus). To appreciate the inherent difficulty of a social dilemma, consider the case of taxation. The existence of government services like hospitals, roads, schools, fire and police departments, well-functioning markets, law courts, and the rule of law itself depends (in almost all countries) on tax revenues, and as much as we might complain about government inefficiency, no society has ever survived for long in the absence of at least some of these major public services. Paying taxes then is clearly to everyone's benefit, to the extent that we'd be crazy *not* to pay them. Yet as Glance and Huberman point out, in not one country in the world is the payment of taxes a voluntary exercise.

Can we not be trusted to do even those things that are obviously in our (collective) best interests? According to the *tragedy of the commons*, an influential theory proposed in the 1970s by the political scientist Garret Hardin, the answer appears to be *no*. Picture a village arranged in a preindustrial style around a large, central, shared plot of land called a *commons*. The villagers use this land mostly for grazing sheep and cattle, which they subsequently shear, milk, or slaughter for their own sustenance or profit. Because the commons isn't owned or governed by anyone, it is free for all to use, but the profits generated by grazing one additional sheep or cow on it go exclusively to the villager who owns the animal. Everyone therefore has an incentive to keep adding to their herds or flocks, thus generating greater and greater individual profits without increasing their overheads.

You can see where this is going. Eventually the commons gets overgrazed to the extent that it can no longer support any livestock at all, and *everyone's* livelihood is undermined. If only the villagers had acted with moderation, there would never have been a problem—the commons would have been able to sustain itself, and the people would have had enough to live off permanently. But even assuming that some village somewhere manages to stumble onto this utopian balance, it is inherently unstable. Even when everyone is happily doing the right thing, each self-interested villager (and they are all self-interested) *always* has an incentive to add one more animal to his flock. No one is going to stop him, and no one is going to complain. It isn't going to cost him anything, and he will be richer as a result or better able to feed his family. The commons isn't going anywhere, and no one is going to notice one more sheep in such a huge and rich pasture, so why not?

Why not indeed. And this is the tragedy, in the Shakespearean sense of inevitable demise. No one is doing anything crazy. Indeed, given what they know about the world, they would be silly (or at least irrational) to do anything different. No matter how ominously disaster looms, the players keep marching in lockstep down the path of destruction, drawn inexorably by their individual self-interest to their collective doom. As its name suggests, Hardin's theory presents a grim view of the world but one that is hard to ignore, reminiscent as it is of so many real-world tragedies—pointless wars prolonged, despicable customs perpetuated, and irreplaceable environments eroded. As much as we would wish these things away if we could, the sad fact is that they are outcomes wrought of our own volition. Like the diner's dilemma, the tragedy of the commons expresses the inescapable conundrum of individuals who have their own interests at heart and who can control only their own decisions, but who have to live with the consequences of everyone else's decisions as well.

INFORMATION CASCADES

But not every dilemma has to end in tears. Just as cultural fads can sweep through a routinely indifferent population, so can social norms and institutions change, sometimes seemingly overnight. As commonplace as it seems today, recycling household materials from plastic bottles to newspapers is a relatively recent phenomenon. In less than a generation, much of the Western industrialized world altered its daily patterns of behavior in response to a distant environmental threat that previously had been perceived as important only to a handful of long-haired tree huggers. How did recycling emerge from its place on the margins of mainstream society to become so entrenched in our expectations of ourselves that despite its continued inconvenience, we no longer question its appropriateness?

Perhaps it's just because recycling a few cans every now and then isn't much of an inconvenience, so the change in habits was relatively costless. But sudden social change can happen even when the individual stakes are much higher, as the citizens of Leipzig demonstrated over thirteen sensational weeks in 1989, when they took to the streets each Monday—at first in thousands, then tens of thousands, and then hundreds of thousands—to protest their oppression under the communist regime of what was then East Germany. Although they are little remembered now, the Leipzig parades probably qualify as a true turning point in history. Not only did they succeed in toppling the East German Socialist Party, but also they led to the fall, only three weeks later, of the Berlin Wall and ultimately the reunification of Germany. Along with so many everyday revolutionaries before them, the Leipzig marchers demonstrated that cooperative unselfish behavior can emerge spontaneously among ordinary people, even when the potential costs—imprisonment, physical harm, and possibly death—are extraordinary. By the end of 1989, in fact, Leipzig had become known throughout East Germany as *Heldenstadt*—city of heroes.

How is it then that even the most rigidly enforced order, the most intransigent dilemma, can suddenly and dramatically break? And if even the most entrenched status quo can fall apart so unexpectedly, how does it manage to maintain itself at all other times, in the face of continual shocks, noise, and disruptions that may be every bit as determined as the ones that ultimately bring it down? Like many researchers before me, I was fascinated by the origins and preconditions of cooperation as a problem in its own right. But what dawned on me slowly, as I ploughed my way through paper after paper in my office in Santa Fe, or wandered the streets around MIT in search of a decent coffee shop, was that all the problems I was reading about, from cultural fads and financial bubbles to sudden outbreaks of cooperation, were different manifestations of *the same problem*.

In the dry language of economics this problem is called an *information cascade*. During such an event, individuals in a population essentially stop behaving like individuals and start to act more like a coherent mass. Sometimes information cascades occur rapidly—the Leipzig parades incubated and exploded within a matter of weeks. And sometimes they happen slowly—new societal norms, like racial equality, women's suffrage, and tolerance of homosexuality, for example, can require generations to become universal. What all information cascades have in common, however, is that once one commences, it becomes self-perpetuating; that is, it picks up new adherents largely on the strength of having attracted previous ones. Hence, an initial shock can propagate throughout a very large system, even when the shock itself is small.

Because they are often of a spectacular or consequential nature, cascades of one sort or another tend to make newsworthy events. While quite understandable, this predilection with action disguises the fact the cascades actually happen rather rarely. The people of East Germany certainly had plenty of reasons to be unhappy with their rulers, but they had been consistently unhappy for thirty years, and only at one other time (in 1953) did any noticeable uprising occur. For every day on which a soccer crowd decimates a stadium, or the stock market decimates itself, there are a thousand days when it doesn't. And for every *Harry Potter* and *Blair Witch Project* that explodes out of nowhere to capture the public's attention, there are thousands of books, movies, authors, and actors who live their entire inconspicuous lives beneath the featureless sea of noise that is modern popular culture. So if we want to understand information cascades, we must account for not only how small shocks can occasionally alter entire systems, but also how most of the time they don't.

It's important to understand that on the surface of things, the various manifestations of information cascades—cultural fads, financial bubbles, and political revolutions, for example—look quite different. To get to the fundamental similarities, one has to strip away the particulars of the circumstances and slog through a thicket of incompatible languages, conflicting terminology, and often opaque technicalities. But a common thread does exist, and after I stared at one problem after another for several months, its rough outline began to coalesce in my mind, like the image in a Chuck Close portrait emerging from the dots as one steps back from the wall. It was an elusive picture, however, and one that required piecing together ideas from economics, game theory, and even experimental psychology.

INFORMATION EXTERNALITIES

IN THE 1950s THE SOCIAL PSYCHOLOGIST SOLOMON ASCH, WHO was none other than Stanley Milgram's mentor, conducted a fascinating series of experiments. Placing groups of eight people together in a room that resembled a small movie theater, Asch's associates projected before them a series of twelve slides displaying vertical line segments of various lengths, much like the picture in Figure 7.1. While flashing the

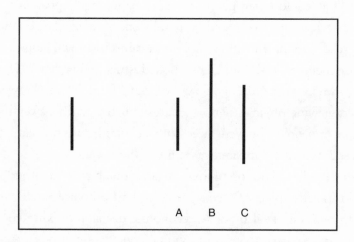

Figure 7.1. Illustration of the test question used by Solomon Asch in his experiments on human decision making in the face of group pressure. The corresponding question would have been, "Which of the three lines on the right is closest in length to the line on the left?" The correct answer was designed to be obvious (here it is A), but seven out of eight subjects were instructed to answer incorrectly (say, B).

slides, the experimenters posed a simple question, which the subjects were to answer in turn, such as "Which of the three lines on the right is closest in length to the line on the left?" The questions were designed so that the answer would be obvious (in Figure 7.1 the answer is clearly A), but the trick was that every member of the audience, bar one, was a plant who had been instructed in advance to give the same *wrong* answer (B, for example).

This setup created an incredibly confusing situation for the poor subject. I know this because when I was giving a talk about information cascades at Yale, a member of the audience—now a distinguished member of the economics faculty—spoke up to say that he had actually been one of Asch's subjects when he was an undergraduate at Princeton. On the one hand, the subject could see with his own eyes and as plain as day that A was closer in length to the line on the left than B was. Yet here were seven people as sensible and rational as he, confidently claiming that the answer was B. How could seven people possibly be wrong? Apparently many subjects decided that they couldn't. In fully one-third of all cases, the subjects agreed with the unanimous opinion, against their own judgment (my audience member, it should be mentioned, stuck to his guns). The subjects' common sense didn't go down without a fight, however. Asch also reported viewing obvious signs of distress, such as visible sweating and agitation, in people who were faced with a choice between violating their own certainty and the apparent certainty of their peers.

But why should our opinions depend so much on what other people think in the first place? Once again, standard economics tells us they should not. The usual model of economic decision making contends that each alternative that an individual considers can be expected to yield a certain payoff or *utility*, which depends partly on the preferences of the individual. Hence, two people with the same preferences will like and dislike the same things, but if preferences are allowed to vary, some people may like the very things that others dislike. Nevertheless, it is always perfectly clear how much any one person wants any one thing, and all that remains to be determined is whether or not they can get it.

This, by the way, is what markets do: they set prices in just such a way that the supply of goods and services exactly matches the prevailing demand, thus ensuring that everyone gets what they want to the extent that they are willing (or able) to pay for it. If a lot of people want the same thing, its price goes up, perhaps so much that some people no longer want it as much as they want something else (like their money, for instance). But, and this is the important bit, the desires of others don't actually cause us to *want* something any more or less, nor do they change its usefulness to us. The preferences remain fixed. All the market determines is the price at which they can be fulfilled. In games of strategy, it gets more complicated. Players now have to take the preferences of others into account when devising their plan of action—hence, what I choose to *do* may be affected by what I know you want. But what I *want* still doesn't change. In such a hyperrational world, there is no point in asking your friends what they think about something, because there is nothing they can tell you that you don't already know. Their preferences can't affect yours.

Back in the real world, however, many problems that we encounter are either too complicated or too uncertain for us to evaluate which alternative is the best. Sometimes, say, when choosing whether or not to adopt a complicated new technology or hire a particular job candidate, we simply lack adequate information about the competing options. And at other times (as with the stock market), we may have access to an abundance of information but lack the ability to process it effectively. Imagine you are walking down the street in a foreign city, looking for a place to eat, and you see two restaurants, side by side, with similar-looking (and equally unfamiliar) menus, indistinguishable prices, and much the same décor. But one is bustling and the other is deserted. Which one do you pick? Unless you have a specific problem with crowds, or you feel sorry for the beckoning waiter in the empty restaurant, you do what we all do in the absence of any better information—you go with the crowd. After all, how could so many people be wrong?

Although Asch's findings make it seem like some kind of flaw, pay-

ing attention to the actions and advice of peers is often a reliable strategy for doing at least reasonably well in a complex and unpredictable
world. From choosing graduate programs to movies, many people consistently opt to minimize potential risk, whether to their career prospects
or their evening's entertainment, by observing and emulating the actions
of others. Even where we explicitly eschew the majority, we are rarely
behaving entirely as atomistic individuals or as pure contrarians.
Rather, we typically have in mind some minority group of actors whom
we *do* wish to emulate. The difference between conventional and uncon-
ventional social actors, therefore, is not so much that the unconven-
tional ones don't pay attention to others—it's just that they are not the
same others as for the conventional actors.

So what Asch found in his experiments appears to be a deeply seated
problem-solving mechanism, an understanding of which requires a
slightly modified view of human rationality compared to what economists traditionally use. Pure economic rationality, after all, makes some
fairly outrageous assumptions about the capabilities of human actors.
Strategic rational actors, for example, are assumed to know everything
about their own preferences and also the preferences of everyone else.
Furthermore, each actor knows that every other actor knows this, and
knows that she knows that he knows that she knows, and so on. Having
achieved this infinite regress of everyone knowing everything, rational
actors are then assumed to act in a manner that optimizes their expected
utility, conditioned on everyone else doing the same.

Of course, not even economists believe that people are really this
smart. Rather they are supposed to act *as if* they are. The standard reasoning is much like the argument given earlier against the existence of
speculative investors: anyone *not* acting in accordance with rational
expectations will do worse than someone who is. So whether the different strategies are intentional or not, people will *learn* to behave rationally simply because that will be observed to work better. Hence,
the only set of actions that matters is the one derived from rational
expectations, because that is the equilibrium to which the system will

inevitably converge. As far as idealized theories of human behavior go, this one has a certain elegant appeal. In fact, from a purely aesthetic point of view, much of neoclassical economics is quite beautiful. But as we saw in the case of speculators, the world it describes often doesn't look much like the real one.

In the 1950s, Herbert Simon, whose ideas we encountered in chapter 4 in the context of preferential growth models, pointed out that as mathematically attractive as it might be, rational utility maximization is ultimately a made-up theory, and so can only be considered a good description of human behavior to the extent that it actually works. If empirical evidence, not to mention common sense, suggests that people do not behave rationally, then why not make up a theory that is more plausible? Replacing mathematical convenience with intuition, Simon proposed that people *try* to behave rationally, but their capacity to do so is bounded by cognitive constraints and limited access to information. In short, they exhibit what he called *bounded rationality*.

Asch's observations of human decision making, therefore, are best understood as a particular manifestation of bounded rationality. Because we are so often uncertain about the best course of action to take, and because we usually lack the capacity to figure it out on our own, we have become conditioned to pay attention to one another, on the assumption that other people know things we don't. We do this so routinely and it works reasonably well so often that we apparently exhibit a reflexive tendency to place considerable weight on the actions of others, even when the answer *is* obvious.

Whenever a person's economic activity is affected by anything outside of the transaction itself, economists call it an *externality*. In general, economics treats externalities as if they are an inconvenient exception to the rule of pure market interactions. But if Asch's results are to be taken seriously, and if everyday experience, from following the crowd in a train station to choosing a cellular phone service, is to be believed, then it appears that what we might call *decision externalities* are everywhere. And in cases like Asch's experiments, where the externalities arise from constraints on what we can know about the world and how we can process even what we do know, we will call them, for want of a better term, *information externalities.*

COERCIVE EXTERNALITIES

ALTHOUGH THE ACTUAL OPINIONS OF ASCH'S SUBJECTS WERE clearly influenced by the (faked) opinions of their peers, some subjects simply felt pressured to indicate their consent, even though *privately* their opinion did not change. As Asch later demonstrated, this perceived pressure to conform was not illusory: in a variant of the experiment, where only a single player was instructed to give the wrong answer, the unwitting majority actually laughed at him. *Coercive externalities,* therefore, may arise in many of the same decision-making scenarios as information externalities and sometimes may be hard to distinguish. Patterns of gang-style crime, for example, are often explained in terms of vulnerable teenagers pressured by their peers and role models into violent or subversive acts in order to prove themselves worthy of membership. But even here, information plays a role. If a young man's dominant examples of economic and social success are those associated with gang leadership, then it is at least plausible that his decision to follow that route, even at the cost of committing crimes he might otherwise perceive as wrong, might seem quite natural and not coerced at all.

Altering one's beliefs in response to the expressed beliefs of others is not restricted to the vulnerable or the ill-informed either. In a groundbreaking study in West Germany in the 1960s and 1970s, the political scientist Elisabeth Noelle-Neumann showed that prior to two national elections, conversations concerning politics displayed a consistent pattern of the holders of the perceived majority opinion growing increasingly vocal and insistent at the expense of the perceived minority. The key word here, however, is *perceived.* As Noelle-Neumann showed, the levels of support for the two political parties, expressed privately by individual citizens, remained roughly constant. What changed was the individuals' perception of the majority opinion, and therefore their expectations of which party would win. In what Noelle-Neumann dubbed a *spiral of silence,* the "minority" became less and less willing to speak their mind publicly, thus reinforcing their minority status and further undermining their willingness to speak up.

Voting, however, is a private activity, so perhaps the balance of pre-election discourse is unimportant. Not so, Noelle-Neumann discovered. Her most striking finding was that on election day, the strongest predictor of electoral success was not which party an individual privately supported but which party he or she expected would win. Beliefs concerning the beliefs of others, therefore, seem capable of influencing individual decision making, even in the privacy of the electoral booth (or possibly by affecting the decision of whether or not even to vote). As with Asch's experiments, and also speculations about the spread of crime, it is somewhat unclear what forces are driving the spiral of silence or influencing an individual's ultimate voting decision, but probably both coercive and information externalities are at play. And decision externalities can arise in other ways as well.

MARKET EXTERNALITIES

STIMULATED BY THE HIGH-TECHNOLOGY BOOM THAT STARTED IN the 1970s, economists have become interested in products that become more valuable as the number of people using them increases. A fax machine, for example, is a self-contained piece of equipment that, like an automobile or a photocopier, has a well-specified set of characteristics. But unlike automobiles and photocopiers, its usefulness depends critically on other people having fax machines also. Unless you place a premium on being the first person to own the latest device, then there is simply no point in buying a fax machine before anyone else. As more and more people buy them, however, they become more and more useful, eventually transforming from a technological curiosity into a virtual necessity.

Because products like the fax machine derive at least some of their individual utility from the presence of other devices that are external to them, the decision to buy one exhibits externalities. But the decision externalities associated with buying a fax machine are not the same as either the cognitive or coercive externalities in Asch's experiments. Although, when choosing which *particular* machine to actually buy, we may rely on the advice of our technically minded friends (and thus make use of information externalities), the decision of whether or not to buy a fax *at all* is essentially an economic calculation that depends only on cost and utility. With products like the fax machine, we may therefore talk of *market externalities* to capture the sense that the utility of a product itself—and often its cost, which tends to decline as the technology becomes more widespread—depends on the number of units sold; hence the size of the market. (Economists, by the way, prefer the term *network externalities,* but because *all* of our decision externalities depend on networks of influences, the term *market externalities* is less ambiguous.)

Market externalities are often reinforced indirectly by what economists call *complementarities.* Two products (or services) are complementary in the sense that each one increases the stand-alone value of the other. Software applications and operating systems, for example, are complementary products because each one is essentially worthless without the other. Market externalities, especially when augmented by complementarities, are capable of generating a positive feedback effect called *increasing returns,* which is similar to the Matthew effect discussed in chapter 4. The more computers that are shipped with a particular operating system, the greater the demand for software applications that will run on them. And the more software that is available for one kind of operating system, the greater the demand for computers that run it. This, in fact, is more or less how the personal computing market became locked onto Microsoft Windows. Because Microsoft got an early lead in the operating system (OS) market (as IBM's preferred choice), and because it naturally had an advantage over everyone else in producing software compatible with its own operating system, it was able to lock in huge shares of *both* the OS *and* the applications market. Apple, by contrast, has always had to struggle with the reality that it controls a relatively small share of the OS market; hence, Macintosh users have never had the same choice of software applications as their Windows counterparts.

COORDINATION EXTERNALITIES

DECISION EXTERNALITIES, THEREFORE, CAN ARISE BECAUSE THE uncertainties of the real world drive us to seek information or advice from our peers (*information externalities*) or even to succumb directly to the pressure they exert on us (*coercive externalities*). Externalities can also arise in the absence of uncertainty, simply because the object of the decision itself is subject to increasing returns (*market externalities*). But there is yet another distinct class of decision externalities that arises out of the structure of public goods games like the diner's dilemma and the tragedy of the commons.

Remember that the way these games work is that "doing the right thing"—whether recycling your plastic and glass, choosing not to double-park on a busy street (even "just for a minute"), or filling up the coffee urn after taking the last cup—is individually costly but collectively beneficial. From the collective point of view, if enough people do the right thing, then everyone is better off—the world doesn't run out of natural resources, the street doesn't get jammed, and the urn is never empty. But from an individual's point of view, if everyone else is doing the right thing, it is always tempting to free ride on their efforts, reaping the benefits of a public resource without contributing to it. Or even worse, if no one is doing the right thing, then what's the point of even trying? It will still cost you the same amount of effort but won't benefit anyone.

The essence of the dilemma is that it is the individuals who are making the decisions, not the collective. Therefore, most strategies for coping with social dilemmas attempt to stack the deck in such a way that individuals have selfish incentives to do what is collectively the desirable thing. Governments do this by fiat, passing civic-minded laws and then coercing their citizens into obedience through the use of legal force. Markets, for their part, also resolve the dilemma, but in a quite different manner. By allocating everything to private ownership and allowing owners to trade their possessions freely, markets (as Adam Smith originally pointed out) are able to harness individual selfishness for the greater good.

But not everything can be regulated effectively by the government or parceled out in easily tradeable packages. Nor would we necessarily want it to be. Absent a world government with the power to subjugate all nations, and short of going to war, there is really no such thing as an enforceable international treaty (one cannot simply put an entire country in jail if it refuses to cooperate). And because many international agreements concern entities that are inherently indivisible, like the atmosphere and the oceans, it is frequently impossible to reconcile individual and collective interests solely through market forces. Rather, international agreements must be reached and maintained through cooperation between independent sovereign states, each of which brings its own logic and interests to the negotiating table. Even punishing an errant nation with, say, trade sanctions requires *other* nations to cooperate by not violating the sanctions for their own gain.

As difficult as collective cooperation can be to generate and sustain in the absence of an effective central government or well-functioning markets, it can and does happen, not only in the international arena but also at the community, company, and household level. Although the conditions required for cooperation to emerge successfully among selfish decision makers are still a matter of debate, a torrent of theoretical and empirical work over the past two decades has shed considerable light on the matter. At the core of all these explanations lie two essential requirements. First, individuals must care about the future. And second, they must believe that their actions affect the decisions of others. If you don't give a damn what happens to you or anyone else one moment from now, then truly you have no incentive to behave any way other than selfishly. Only if the future matters can the prospect of doing unto others make a short-term sacrifice seem worthwhile. Caring about the future, however, is not enough. Only if you believe that by supporting the collective interest you will cause others to join you, does the future give you any selfish incentive to do so. And the only way you can assess how much of a difference you can make, and whether it will be enough, is if you pay attention to the actions of others. If it looks like enough people are joining in, you may decide that it is worth joining in also. If not, you won't. As a consequence, the decision of whether or not to cooperate depends critically on what we will call *coordination externalities*.

SOCIAL DECISION MAKING

Whether compensating for lack of information, succumbing to peer pressure, harnessing the benefits of a shared technology, or attempting to coordinate our common interests, we humans continuously, naturally, inevitably, and often unconsciously pay attention to each other when making all manner of decisions, from the trivial to the life changing. This, however, is not a view of ourselves with which we are entirely comfortable. We like to think of ourselves as individuals, capable of making up our own minds about what we think is important, and how to live our lives. Particularly in the United States, the cult of the individual has gained a large and devoted following, governing both our intuitions and our institutions. Individuals are to be seen as independent entities, their decisions are to be treated as originating from within, and the outcomes they experience are to be considered indicators of their innate qualities and talents.

It's a nice story, implying as it does not only the theoretically attractive notion that individuals can be modeled as rationally optimizing agents, but also the morally appealing message that each person is responsible for his or her own actions. However, there is a difference between holding someone accountable for their actions and believing the *explanation* for those actions is entirely self-contained. Whether we are aware of it or not, we rarely, if ever, make decisions completely independently and in isolation. Often we are conditioned by our circumstances, our particular life histories, and our culture. We also cannot help but be influenced by the mesmerizing pool of universally available, often media-driven information in which we continually swim. In determining the kind of person that we are and the background picture against which our lives play out, these generic influences determine both the expertise and the preferences that we bring to any decision-making scenario. But once we are in the scenario, even our experience and predispositions may be insufficient to sway us entirely one way or the other. This is where externalities—whether information, coercive, market, or coordination externalities—enter to play a crucial role. When push comes to shove, humans are fundamentally social creatures, and to ignore the role of social information in human decision making—to ignore the role of externalities—is to misconstrue the process by which we come to do the things we do.

I recently read an article in the newspaper about the rising popularity of body piercing, especially among teenagers. According to the defiant youngsters interviewed, they had not made the decision to get pierced in order to drive their fuddy-duddy parents crazy, or even because their friends were doing it, but solely for their own satisfaction—according to one young woman, "because I wanted it." Well maybe so, but this glib explanation really just raises the question, why did she want it? The young lady in the article would no doubt claim that it was an independent choice, independence being an especially sought-after commodity among American teenagers. But the striking temporal, geographical, and social clustering of these "independent" piercing decisions suggests that they are anything but that. Rather, the trend has flared like a contagion, spreading from city to city and across social groups in a cascade of increasingly contingent decisions, each of which is made by an individual unaware of the larger pattern into which his or her choice fits. But the pattern is there nonetheless, and it is a pattern shared by a multitude of other social phenomena, ranging from the dizzy heights of modern finance to the gritty depths of grass-roots rebellion. To understand the pattern, however, we need to delve further into the rules by which individuals make decisions, and how, in the process, our apparently independent choices become inextricably bound together.

CHAPTER EIGHT

Thresholds, Cascades, and Predictability

I REMEMBER TELLING STEVE ABOUT INFORMATION CASCADES AT the 2000 AAAS conference in Washington, D.C., where Harrison White gave his talk about social contexts that started us off on our affiliation networks project with Mark. As we wandered around the National Zoo on a chilly Sunday morning, waiting for the monkeys to wake up, we agreed that one of the most intriguing features of the cascade problem was how most of the time the system is completely stable, even in the face of frequent external shocks. But once in a while, for reasons that are never obvious beforehand, one such shock gets blown out of all proportion in the form of a cascade.

And the key to a cascade, it seemed, is that when making decisions about how to act or what to buy, individuals are influenced not only by their own pasts, perceptions, and prejudices but also by each other. So only by understanding the dynamics of decisions *with externalities* can collective behavior, from fads to financial bubbles, be understood. Lurking once again in the guts of the problem is the network—that ubiquitous web of signals and interactions through which the influence of one person passes to another. Between the two of us, Steve and I had thought a good deal about contagious entities spreading on networks, but mostly we had imagined ourselves to be talking about biological diseases like HIV and Ebola, or computer viruses. We had done some work, as part of my Ph.D. research, on the evolution of cooperation in small-world networks and on a special case of what is called a *voter model* (similar to the spiral-of-silence problem studied by Noelle-Neumann). But at the time, we hadn't thought of either of these problems as related to contagion.

Now it seemed clear that contagion in a network was every bit as central to the outbreak of cooperation or the bursting of a market bubble as it was to an epidemic of disease. It just wasn't the same kind of contagion. This point is particularly important because typically when we talk about social contagion problems, we use the language of disease. Thus, we speak of ideas as *infectious,* crime waves as *epidemics,* and market safeguards as building *immunity* against financial distress. As metaphors, there is nothing wrong with these descriptors—after all, they are part of the lexicon and often convey the general point vividly. But the metaphors can be misleading because they also suggest that ideas spread from person to person in the same way that diseases do—that all kinds of contagion are essentially the same. They are not, as we can understand by thinking again about the psychology of decision making.

THRESHOLD MODELS OF DECISIONS

IMAGINE YOURSELF TO BE IN ONE OF SOLOMON ASCH'S EXPERIMENTS with seven other people, some of whom have been told to give the right answer, A, and others of whom are deliberately giving the wrong answer, B. You don't know this, but at first it doesn't matter because as soon as you see the slide, you're pretty confident that the answer is A. Before you get to voice your opinion, however, you have to wait for everyone else to answer, during which time you might change your mind. Imagine that six of the seven people vote for A, thus reinforcing your opinion, and one person votes for B. Obviously that person is an

idiot and everybody is laughing at him—there's no way you are changing your mind. If two people vote B, probably nothing would be different—your natural opinion is still reinforced by a large majority, so you
have no reason to doubt yourself. If three or four people vote for B,
however, you might start to worry. What is going on? How can a group
of people be so dramatically divided over something so obvious? What
are you missing? Maybe you're not so sure after all, and if you're the
kind of person who is plagued by self-doubt, you might change your
mind. But maybe you're *really* confident of the answer, so still you are
unswayed. Okay, so now five people vote B, or six, or all seven!

At what point do you crack? At what point do you mentally throw
up your hands and concede that you don't understand whatever it is
that everyone else is understanding? Maybe never. Some people never
change their minds, but in situations where we harbor even the tiniest
bit of uncertainty, many of us do. Certainly that's what Asch's experiments indicated. And a closer look at his results reveals an even more
interesting story. By varying the number of people in the room, Asch
showed that the tendency of his subjects to agree with the majority
opinion was largely independent of the absolute number. It didn't matter whether three people or eight people gave a particular answer—only
that their opinion was unanimous. The second thing that Asch noticed
was that if even a small crack appeared in the wall of unanimity—if
a single member of the majority was instructed to give the correct
answer, thus agreeing with the subject—the subject's confidence would
often reassert itself, resulting in a steep drop-off in the error rate.

These variations on Asch's main result reveal some important subtleties to the general rule that social beings pay attention to one another
when making up their minds. First, it is not so much the absolute number of people making a particular choice that compels one to follow
suit, so much as the relative, or *fractional,* number choosing one alternative over another. This is not to say that the sample size is irrelevant.
If you only solicit a few opinions before making a decision, then each
individual opinion carries more weight than if you sample many peo-

ple. But once the size of your neighborhood is set, and once the choice—option A versus option B—is presented, it is the *relative* number of your neighbors choosing A over B that drives your decision. Second, even small changes in the fraction of your neighbors making one choice versus the other can have a dramatic effect on your ultimate decision. The first time we hear an apocryphal rumor, for example, we might be disinclined to believe it. But if we hear that same rumor from a second and possibly a third source, then *at some point* our tendency is to switch from skepticism to (possibly grudging) acceptance. Again, how could so many people be mistaken?

So although making a decision can be thought of as getting "infected" by a particular idea, the *mechanism* of the contagion is very different from the contagion of a disease. In disease, exposure to a single infected neighbor carries the same probability of infection regardless of how many unsuccessful exposures occurred beforehand. Disease contagion events, in other words, occur *independently* of one another. In the case of a sexually transmitted disease, for example, if someone engages in sex with an infected partner and is lucky enough not to get infected, the next time that person is exposed, he or she is neither less nor more likely to escape harm—each time is simply an independent roll of the dice. A graph representing the cumulative likelihood of infection is shown in Figure 8.1. Although it flattens out for large numbers of infected neighbors, for small numbers each additional exposure increases the overall probablity of infection by roughly the same amount.

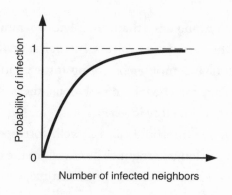

Figure 8.1. Probability of infection in the standard disease-spreading model as a function of the number of an individual's infected neighbors.

Social contagion, by contrast, is a highly contingent process, the impact of a particular person's opinion depending, possibly dramatically, on the other opinions solicited. A negative opinion of a potential job candidate, for example, might be the kiss of death if it comes on the back of prior negative remarks, or it may be discounted entirely if followed by a slew of positive reports. A social decision rule, therefore, looks something like Figure 8.2, where the probability of choosing outcome A increases at first very slowly with the fraction of neighbors choosing A, before jumping rapidly once a *critical threshold* has been exceeded. Because of this characteristically sudden switch from one alternative to the other, we call this class of decision rule a *threshold rule*, where the position of a person's threshold signifies how easily he or she is influenced. In Asch's experiments, the threshold would lie

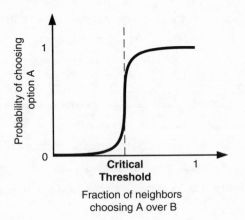

Figure 8.2. Probability of choosing A versus B in social decision making, as a function of the fraction of neighbors choosing A. As the individual's *critical threshold* is reached, the probability of choosing A jumps rapidly from near zero to near one.

very close to one, because anything less than complete unanimity resulted in very few errors on the part of the subjects. But in less certain scenarios, such as choosing a new computer or voting for a political party, where the better alternative may be far from obvious, the corresponding thresholds might be considerably lower.

There are other ways to derive threshold rules as well, corresponding to the different kinds of decision externalities discussed in the last chapter. The decision to adopt a new technology, for example, can be represented by a threshold rule if the technology is subject to market externalities. It doesn't matter that the origin of the externalities is completely different from that of the information externalities in Asch's experiments. In the example of the fax machine, as far as the purchasing decision is concerned, all that matters (besides the cost) is that a certain fraction of the people you communicate with (or would like to communicate with) have fax machines. Furthermore, the probability of adopting such a technology can change rapidly, if the population of owners, as a fraction of the total population with which you communicate, increases past the threshold at which the purchase makes economic sense.

Threshold rules can also be derived from coordination externalities in social dilemmas, where the individual cost of contribution to some public good is worthwhile only as long as enough other people also contribute. The actual position of an individual's threshold depends on precisely to what extent that individual cares about future payoffs versus short-term gain from acting selfishly, and also how much influence he or she perceives themselves as having. It's possible for individuals to have such a high threshold that they never contribute, no matter what other people do, or such a low threshold that they always contribute. The important point is that regardless of what it is or how specifically it got there, everyone has *some* threshold.

And that is why threshold models of decision making are so important to understand. While there may be many ways to derive a threshold rule—whether from the logic of game theory, the mathematics of increasing returns, or experimental observation—once its existence is established, we no longer need to concern ourselves with how it was derived. Because we are interested in *collective* decision making, all we need to know about the decision rule itself is that it captures some essential features of *individual* decision making. What concerns us now is the *consequences* at the *population* level. In other words, when everyone is casting around for signals of what to do and giving out signals themselves, on what sort of decision is the population *as a whole* likely to converge? Will cooperation emerge or will the status quo remain? Will a cascade of buying drive prices into an unstable bubble or will a sense of intrinsic value prevail? Will a technological innovation succeed or fail? These are the kinds of questions that simple models based on threshold rules can hope to answer. And because the threshold rule is representative of so many social decision-making scenarios, whatever it tells us about collective decision making should apply regardless of many of the details.

CAPTURING DIFFERENCES

ONCE AGAIN, HOWEVER, SOME DETAILS DO MATTER. MOST IMPORTANT, in social contagion problems of all varieties, we need to account for the basic observation that *people are different*. Some people, for whatever reason, are more altruistic than others and are prepared to bear a higher personal cost in order to support a cause that hasn't yet had the chance to succeed. These are the first of the Leipzig marchers, the pro-testers of Tiananmen Square, the Martin Luthers and the Martin Luther Kings—all those who risk their lives and liberties in the course of fighting their crusades, and rarely get to retire to the Hamptons, but who serve the critical role of being the first. Others are sympathetic and willing to contribute, but not until the project seems reasonably likely to succeed and the costs of joining have correspondingly lessened. Still oth-ers will join in only when success seems so certain they are afraid of being left out.

Equally important from a decision-making perspective is that indi-viduals generally possess different levels of information or expertise relevant to a problem; hence, some will be more easily influenced than others. Also, people vary in the strength of their convictions, regardless of whether they are better informed or not. Some people are natural innovators, continually dreaming up new ideas or new uses for existing products. Others, less creative, are constantly on the hunt for the latest gadget or trend, in the hope of profiting from an early investment or simply showing off to their friends. Still others will prefer to stick with what they already understand, no matter how much the world tries to change around them. Most of us, meanwhile, are somewhere in the middle, too busy with our lives to spend much time inventing or scanning for inventions, but happy to jump on the bandwagon once the risk of looking silly appears minimal.

Although the variability of human dispositions and preferences is complicated in real life, it is relatively straightforward to capture in our threshold model. Unlike most models in physics (and even economics) where individuals are generally considered to be identical, here the individuals in our network can have different thresholds, where the overall *distribution of thresholds* (an example of which is shown in Figure 8.3) can be interpreted as a measure of variability across the population as whole. This kind of variability, what we might call *intrinsic variability*, turns out to be important to the propagation of information cascades—

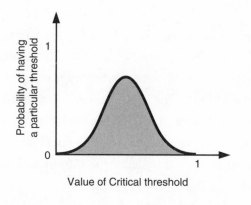

Figure 8.3.
The probability distribution of thresholds over the population captures the variability of individual characteristics.

sometimes in surprising ways. For example, the presence of a wide range of personal thresholds in a population tends to increase the chance of new ideas or products catching on considerably.

Another kind of variability is also important: if it matters so much that we pay attention to one another, then it must matter how *many* others we pay attention to. When I am buying new clothes, for example, I almost always take a female chaperone along, lest the prospect of making unsupervised fashion decisions causes me to panic and run. Ideally, I would like to drag along a few of them, not just because it would do wonders for my image but because multiple opinions would probably yield more reliable information. Typically, however, convincing even one of my female friends to come shopping is hard enough, so one usually has to do. That being the case, I have to choose my companion carefully, as—having no sense of style myself—her opinion is the equivalent of Caesar's thumb, and what I end up wearing is entirely at her mercy. In other situations, from choosing whether or not to see a particular movie, visit a restaurant, buy a new laptop, or hire a job candidate, we may solicit a range of opinions, depending on how important the decision is to us and how much time we have. More opinions, however, are not always better. The more opinions we solicit in making a decision, the less influenced we are by any *one* of them, and therefore the less impact any single good suggestion is likely to have.

One way to think of an aggregate statistic, like opinion polls or the market share of a particular product, is as basically the same kind of socially transmitted information we get from our friends but averaged over a much larger population. Ford frequently promotes its Explorer as "America's number 1 selling SUV," with the implication that if so many other people like it, you will too. The share price of a particular stock is another example: the more people *in the entire market* who want to buy the stock, the higher its price. On the surface, it seems like this kind of global information ought to be more reliable than simply asking your friends, coming as it does from such a large sample size.

Nevertheless, we still tend to be influenced disproportionately by the opinions or actions of our immediate friends, contacts, sources, or coworkers. For example, when deciding whether to buy a Macintosh or a PC laptop, the fact that globally PCs far outsell Macintoshes would seem all but irrelevant if everyone you work with uses a Mac. A recent advertising campaign by Apple, in fact, intimated that if you are an accountant (read "dull, boring, shunned at parties"), you probably use a PC. But if you work in the arts, design, or fashion (implying "hip, edgy, on the list"), then you are very likely to use a Mac. The message is that the information your friends give you is more important than any global information you might receive because the former is more relevant to *you*. So asking too few people is potentially bad, because you make yourself susceptible to errors. But asking too many people is bad also, because the relevant information then gets lost in the noise.

Networks of social information, moreover, are important not just because they help us make better individual decisions but also because they allow things that have caught on in one setting to spill over into another. Since this kind of spillover is critical to the dynamics of a cascade, social networks are central to the notion of a little thing becoming big. When 3Com released its first version of the Palm Pilot, only the most radical aficionados of technology bought them. This small group of people, mostly engineers and technology workers in Silicon Valley and the encompassing Bay Area of northern California, didn't need anyone else's say-so in order to have the newest new thing. What they really cared about was the innovation itself—they just had to have it, regardless of what anyone else thought. But true technophiles, like true hipsters and true acolytes, are relatively rare—rare enough that by themselves, they cannot make a new product successful. If, however, they can make it big enough in their little world so that it spills over into the other little worlds to which they are connected, then all those little worlds together may give the product the push it needs to launch into the larger world and become a cascade. But how must they be connected?

CASCADES IN SOCIAL NETWORKS

THAT WAS THE QUESTION I STARTED OUT ON. ULTIMATELY, I WANTED to figure out what particular features of social networks, like the presence of groups and communities, and the propensity of individuals to connect across them, most favored the growth of a small initial influence into a global movement. If one wanted to start a revolution, for example, or a fad, how ought one to go about seeding it? Did networks have weak points—structural Achilles' heels—such that if they were hit in precisely the right way, a small shock would explode like an epidemic, each successive decision generating the conditions for the next? And if so, could one exploit that knowledge to enhance the likelihood of a cascade? Or alternatively, to prevent one? Could the same reasoning be extended to engineered systems like the power grid, to reduce the probability of a cascading failure like the one of August 1996? Could firewalls, in a sense, be inserted in networks, in much the same way that they are placed in buildings to contain a blaze?

These were all good questions, but as I dug deeper into the problem, it became obvious that the answers weren't going to come easily. Social contagion, it turns out, is even more counterintuitive than biological contagion, because in threshold models, the impact of one person's action on another's depends critically on what other influences the latter has been exposed to. In disease spreading, as already pointed out, we don't have to worry about this effect because every contagion event can be considered independently of any other. But in social contagion, it makes all the difference in the world.

An isolated group of people—a religious cult like the Branch Davidians, for example—can maintain completely implausible beliefs as long as they remain in a context where they can continually reinforce each other and prevent each other from interacting with the outside world. But for this same reason, their ideas tend to remain confined to the particular group in which they originate. At the opposite extreme, individuals who participate simultaneously in many different groups can tell more kinds of people their ideas and likewise can access a broader range of information. But they are less likely to be dominated by any single worldview, and may often have to sell their ideas themselves, with little support from others. The spread of ideas, therefore, unlike the spread of disease, requires a trade-off between cohesion within groups and connectivity across them.

One of the quirky facts about Ithaca that I learned when I was studying at Cornell is that the city supports an alternative currency, called *Ithaca hours,* that can be earned and spent at a number of businesses downtown. As odd as it sounds, this system has persisted stably for over a decade, but it has also remained highly localized, not even diffusing up the hill to the businesses surrounding Cornell's campus. When I left Ithaca in 1997 and first moved to New York (as a postdoc at Columbia), I remember that Citibank and Chase Manhattan were also trying to introduce a certain kind of alternative currency—an electronic cash card—on the Upper West Side of Manhattan. Despite heavy promotion by two of the largest banks in the country, this supposedly superior alternative to paper money completely failed to catch on.

There are many differences between these two examples, but the one that is relevant to this discussion is that in Ithaca, the network of customers and vendors is sufficiently densely connected to be self-sustaining. The Upper West Side, by contrast, is too integrated into the rest of New York for any individual to have enough stake in a purely local alternative to cash. If, however, the cash cards *had* caught on on the Upper West Side, it seems plausible that, unlike Ithaca hours, the innovation would have spread, for precisely the same reason that it failed. Again, the success of an innovation appears to require a trade-off between local reinforcement and global connectivity. And this requirement renders social contagion significantly harder to understand than its biological counterpart, where connectivity is all that matters.

After a good deal of unsuccessful casting around, I eventually conceded that as simple as the threshold model was, it was going to have to get simpler if I was going to sort out the complexities of group structure from the notion of cascades propagating through a network of associations. So I decided to start with a network that didn't have any group structure in it at all: a random graph. Although random graphs are not particularly good models of real social networks, they are nevertheless a good place to start. I promised myself that as long as I didn't *stop* at random graphs, it would be okay to use them as a launch pad from which to explore more realistic representations of networks. As we will see, matters get quite complicated enough even with random graphs, but we can still learn some surprisingly general lessons.

Because the technical version of the threshold model is a little abstract, it helps to use the intuitive terminology of the *diffusion of innovations,* introduced in the 1960s by Everett Rogers. Although the word *innovation* is often associated with the introduction of new technologies, the concept can be used to refer to ideas and practices as well. An innovation, therefore, can be quite profound, like a revolutionary new idea or a new social norm that will last for generations, or it can be quite banal, like a scooter or a fashion item that will be popular for only a season. It also can be virtually anything in between, including new medical drugs, new manufacturing technologies, new management theories, and new electronic devices. Correspondingly, the term *innovators* can be used to refer not only to individuals who introduce new devices but also to advocates of new ideas, or more generally still, any small shock that disturbs a previously quiescent system. And the phrase *early adopters,* often used to describe the individuals who immediately seize upon a new product or service and advocate it to others, covers all acolytes, apostles, and followers of revolutionaries as well. Early adopters are simply the members of a population who, like the Silicon Valley techies from before, are the first to be influenced by an external stimulus.

As evocative as Rogers' terms are, however, they aren't precise enough to avoid ambiguity. For example, it can be hard to tell if individuals have adopted a new idea because they were inherently predisposed to it (they had a low threshold) or because they were subject to very strong external influences (their particluar neighborhood happened to contain a high density of previous adopters). Either explanation could account for an early adoption, but each carries a very different implication for the individuals in question. Most of the time, we simply accept that terms like *innovators* and *early adopters* have subjective meanings, and we use them in whatever way suits our purpose at the time. But here, because we have a precise, mathematical framework with which to work, we can do a little better. And if we are to make any progress, we will need to.

So from now on, the term *innovator* refers to a node that is activated randomly at the beginning of an *innovation cycle*. When the cycle commences, every node is considered *inactive (off);* then the innovation is triggered by selecting one or more nodes (constituting the initial seed) at random to be *activated* (switched to an *on* state). These are our innovators. The label *early adopter* can now also be defined as a node that will switch from an inactive to an active state *under the influence of a single active neighbor.* Since we want to understand the susceptibility of networks to cascades, we call nodes that are early adopters in this precise sense *vulnerable,* because they can be activated by the smallest possible influence from their network neighbors. All other nodes meanwhile are *stable* (although, as we will see later, even these stable nodes can be activated under the right circumstances). A node, therefore, can be vulnerable in one of two ways: either because it has a low threshold (thus, a predisposition for change); or because it possesses only a very few neighbors, each of which thereby exerts significant influence.

Early adopters, in fact, can have virtually any threshold at all, as long as they have few enough neighbors. This may seem like an odd distinction, but its worth understanding because it changes our entire approach to the problem. Instead of judging early adopters in terms of their thresholds, we can focus on their degree, which, recall from chapter 4, refers to how many neighbors they have. For example, in Figure 8.4, assume that node A has a threshold of one-third. In the top panel, A has three neighbors, one of whom is active. Because this single active node constitutes one-third of A's neighborhood, A's threshold is reached and it activates; hence A behaves like an early adopter. In the bottom panel, by contrast, A has the same threshold, but now it has four neighbors instead of three. Since its single active neighbor now comprises only one-fourth of its total input, A doesn't activate. Depending on its degree, therefore, a threshold of one-third may or may not be low enough to make A an early adopter. Or to put it the other way around, we could say that for a threshold of one-third, A has a *critical upper degree* of three, where critical upper degree is defined as the maximum number of neighbors a node can have and still be activated by any *one* neighbor. If A's threshold were lower (say one-fourth), it would have a higher critical upper degree (four), and vice versa. The important point is that for any given threshold, we can always determine an equivalent critical upper degree. If a node has more neighbors

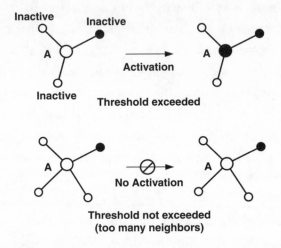

Figure 8.4. For any given threshold a node can only be activated by a single neighbor if its degree is less than or equal to the critical upper degree corresponding to its threshold. Here, node A has a threshold of one-third and therefore a critical upper degree of three. In the top panel, A has three neighbors; hence, it activates. But in the bottom panel, it has four neighbors, so it remains inactive.

than its critical upper degree, then it will be stable with respect to single-neighbor influences, and if not, it will be vulnerable. Variability of degree—our observation from before, that some people have more friends or simply solicit more opinions than others—is therefore central to the stability of individuals and consequently to the dynamics of cascades.

CASCADES AND PERCOLATION

WITH THIS FRAMEWORK, THE QUESTION OF WHETHER OR NOT an information cascade occurs in a population of decision makers can now be specified precisely. In our network of individuals, each one has an internal threshold and a set of network neighbors to whom he or she pays attention. At the start of an innovation cycle, a single innovation is released somewhere in the network, and then one of two things must happen before the cycle ends: either the innovation dies out; or else it explodes into an information cascade.

But how widely must an innovation propagate before it qualifies as a cascade? The key to answering this question turns out to be something we have already encountered: the concept of percolation. Remember that in the context of disease spreading, we defined the condition for an epidemic to be the existence of a single, connected cluster, the *percolating cluster,* which occupies a finite fraction of the network no matter how large the network is. By analogy, when a percolating cluster arises in the context of social contagion, we say the system is susceptible to a *global cascade.* Cascades of smaller sizes happen all the time—every shock, in fact, triggers a cascade of some size, even if just the lonely innovator himself. But only global cascades grow in a truly self-perpetuating manner, thus altering the states of entire systems. So just as we were interested in epidemics of disease, rather than just outbreaks, it is the condition for a *global* cascade that we are now after.

Unlike in disease spreading, however, where every node has the same probability of being a part of an infected cluster, now we have two kinds of nodes—*vulnerable* and *stable*—that we need to consider separately. If we imagine what happens when an innovation is introduced into an initially inactive population, we can see that it can only spread if the initial innovator is connected to at least one early adopter. Obviously the more early adopters there are in the population, the more likely a particular innovation is to spread. And the larger the connected cluster of early adopters in which the innovation lands, the farther it will spread. If the vulnerable cluster that is "hit" by an innovation (that is, the cluster containing an innovator) happens to percolate throughout the network, then the innovation will trigger a global cascade. So if the network contains a *percolating vulnerable cluster*, then it is possible for global cascades to occur, and if it doesn't, then they cannot—they will always die out before activating more than a tiny fraction of the population.

The problem of determining whether or not successful cascades can occur in a system, therefore, reduces to one of showing that a percolating vulnerable cluster does or does not exist. Believe it or not, we have just taken a big step forward. By converting what was originally a dynamic phenomenon (the trajectory of every cascade from a small initial shock to its final state) into a static percolation model (the sizes of vulnerable clusters), we have simplified our task immensely *without* losing the essence of our original inquiry. It's still a hard problem, however. A great deal of progress has been made on percolation models of different kinds in the last thirty years, but as yet, no completely general solution exists. In fact, because percolation has been developed almost entirely by physicists, and because physics applications typically involve regular lattices, very little is known about percolation on more complicated network structures such as social networks.

This is precisely where the extremely simple structure of random graphs comes into its own. In fact, it was at this point in thinking about the problem that I realized I would need to understand cascades on random graphs first. Also around this time Mark, Steve, and I were figuring out the mathematical techniques to compute the connectivity properties of random networks (see chapter 4), which we later modified with the help of Duncan Callaway to study percolation in the context of network robustness (see chapter 6). As luck would have it, the same tools turned out to be almost exactly applicable to the problem of finding percolating vulnerable clusters—but not quite applicable, because now we are dealing with an odd kind of percolation. As Figure 8.4 suggests, nodes with a large number of neighbors tend to be stable with respect to single-neighbor influences, and stable nodes, by definition, cannot be part of any vulnerable cluster. Hence, the vulnerable cluster needs to percolate effectively in the absence of the most-connected nodes in the network. Not surprisingly, this deviation from standard percolation has significant consequences for the results.

Although the mathematical details of the method are quite technical, the main results can be understood rather easily by considering what is called a *phase diagram,* an example of which appears in Figure 8.5. The horizontal axis represents the average value of the threshold distribution—that is, the typical resistance of an individual to a new idea. And the vertical axis is the average number of network neighbors (degree) to whom each individual is paying attention. The phase diagram, therefore, encapsulates all possible systems that can be represented in the simple framework of the model. Each point in the plane represents a particular kind of system, with a specified network density on the one hand, and an average threshold for the population on the other. The lower the average threshold, the more predisposed the population is to change, so one would expect cascades to arise more frequently on the left side of the diagram (where thresholds are low) than on the right side. And indeed, that's what we see. But the relationship is complicated by the presence of the network through which the cascade must propagate.

The reason why Figure 8.5 is called a phase diagram is that the solid black line separates the space of all possible systems into two *phases.* The shaded region inside the line represents one phase of the system in which global cascades can occur. They don't necessarily occur—and that's important—but they can. Outside the line, by contrast, global cascades can never occur. What the distinct boundaries of this *cascade window* tell us then is that there are three ways in which cascades can be forbidden. The first one is obvious: if everyone's threshold is too high, no one will ever change and the system will remain stable regardless of how it is connected. Even when this is not the case, cascades can still be forbidden by the network itself, in two ways: either it is not well connected enough or (and this is the surprising part) it is *too well* connected.

The other important feature of the phase diagram is that near any boundary of the cascade window, the system goes through a *phase transition*. This is a standard feature of most percolation problems. But what makes this kind of percolation different from the kind that we considered in chapter 6 is that the cascade window has two boundaries: an upper boundary, where the network is highly connected, and a lower

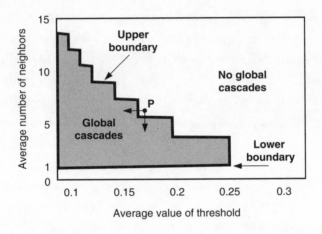

Figure 8.5. The phase diagram of the cascade model. Each point in the plane is equivalent to a particular choice of parameters (the average value of the threshold, and the average number of neighbors, or *degree*). Global cascades can occur within the solid line (the cascade window) but not outside it. The boundaries of the cascade window correspond to phase transitions in the behavior of the system. The point P represents a state of the system for which global cascades are not possible. Starting at P, global cascades can be induced either by lowering the average threshold of the population (left arrow) equivalent to increasing the inherent appeal of the innovation, *or* by reducing the density of the network (down arrow).

boundary, where it is not well connected at all. This feature alone makes cascades different from disease epidemics, where greater connectivity always makes diseases more likely to spread. (If we constructed a phase diagram for disease epidemics, the lower boundary would still be there, but the upper boundary would disappear.) However, the differences are actually greater still. As we will see, the phase transitions that occur at each of the two boundaries are fundamentally different. And by examining the nature of these phase transitions, we can make predictions about what sorts of cascades are possible, how large we should expect them to be, and how often they should occur.

PHASE TRANSITIONS AND CASCADES

At THE LOWER BOUNDARY OF THE CASCADE WINDOW, WHERE THE network is poorly connected, we see a phase transition that is very similar to the one we encountered in chapter 6 for biological models of contagion. The explanation is that when nodes have on average only one neighbor, they are almost always below their critical upper degree, hence vulnerable to new influences, regardless of what their particular threshold is. However, because the network is so disconnected, those influences can never propagate very far. As a result, innovations have a tendency to spread initially but always get contained within the small connected cluster in which they start. Only when the network becomes dense enough do we see the percolating vulnerable cluster appear. But because *most* nodes are still vulnerable in this regime, the percolating vulnerable cluster really is just the same as the giant component of a random graph that we encountered originally in chapter 2 and then again in chapter 6.

Near the lower boundary, therefore, social contagion is largely equivalent to biological contagion, because it undergoes the same phase transition that epidemics of disease do. So under some conditions the conflation of the two kinds of contagion is legitimate after all, in that the differences between the two kinds of models don't appear to affect the outcome. And for the same reason—that network connectivity, rather than the resilience of individual decision makers, is the principal obstacle to a successful cascade—it is also true that in poorly connected networks, highly connected individuals are disproportionately effective in propagating social contagion. This second observation reflects standard diffusion-of-innovations thinking, according to which opinion leaders and centrally situated actors are considered the most effective promoters of a new idea, practice, or technology.

For example, in his recent book, *The Tipping Point*, the writer and journalist Malcolm Gladwell emphasizes the role that highly connected individuals play in social contagion, where his use of the term *tipping point* corresponds roughly to the notion of a global cascade. Although Gladwell develops his ideas about the diffusion of ideas from the premise that social contagion operates no differently from disease contagion, his observations are in general agreement with those of the threshold model, *provided* that the network of decision makers is poorly connected. Gladwell's *connectors* come from that rare breed of socially prodigious individuals who not only maintain superhuman Rolodexes but also span many different social groups. And in a world where most people have only a few friends, or solicit very few opinions when making decisions, it does indeed seem that the occasional connector would occupy a position of great influence.

But influences can also be stymied if a network is too well connected. As discussed earlier, the more people whose actions or opinions you take into account before making a decision, the less influence any *one* of them will have over you. So when *everyone* is paying attention to many others, no single innovator, acting alone, can activate any one of them. This feature of social contagion is what sets it apart from biological contagion, where a susceptible individual's contact with a single infective has the same effect, regardless of how many other contacts the susceptible has had. In social contagion, remember, it is the relative number of "infected" versus "uninfected"—*active* versus *inactive*—neighbors that matters. So although highly connected networks might seem, on the surface, to favor the propagation of all sorts of influence, they do not necessarily support cascades of social influence. Because in such a network, all individuals are locally stable, no cascade can ever get started in the first place.

Networks that are not connected enough, therefore, prohibit global cascades because the cascade has no way of jumping from one vulnerable cluster to another. And networks that are too highly connected prohibit cascades also, but for a different reason: they are locked into a kind of stasis, each node constraining the influence of any other and being constrained itself. So our anecdotal observation from earlier can now be made more precise: in social contagion, a system will only experience global cascades if it strikes a trade-off, specified by the cascade window of Figure 8.5, between local stability and global connectivity.

CROSSING THE CHASM

BUT SOCIAL CONTAGION HAS ANOTHER SURPRISE IN STORE. RIGHT at the upper boundary of the cascade window, the density of vulnerable nodes is just enough for the network to contain a percolating vulnerable cluster. In this precarious state, the system is locally stable almost everywhere except right around the vulnerable cluster itself. And because, just inside the window, the vulnerable cluster only occupies a small fraction of the entire network, the chance that a single innovation will strike it is small. Cascades, therefore, will tend to be very rare, and most of the time the system will behave as if it is not just locally stable but globally stable as well. Once in a while, however—and this could mean one time in a hundred or one in a million—a random innovation will strike the vulnerable cluster, triggering a cascade. So far, this isn't so different from what happens at the lower boundary, where global cascades also happen rarely. But once the cascade gets rolling, the two scenarios diverge rapidly.

Remember that at the lower boundary, the cascade propagates until it occupies the vulnerable cluster and then runs out of places to go; hence, cascades only occupy a relatively small fraction of the entire network. However, at the upper boundary, because the network is so highly connected, the vulnerable cluster of early adopters is tightly integrated into the rest of the network (the *early* and *late majority,* in Rogers's terminology). This much larger population is still stable with respect to individual innovators, but once the entire vulnerable cluster has been activated, these initially stable nodes become exposed to *multiple* early adopters. And the presence of these multiple active influences *is* sufficient to exceed the thresholds of even quite stable nodes, so they start to activate as well.

This event, when it happens, is what the business consultant and writer Geoffrey Moore calls *crossing the chasm,* referring to the leap a successful innovation (like our example of Palm Pilots from before) needs to make from its initial community of early adopters to the much larger general population. At the lower boundary, there is no such chasm to be crossed, just early adopting clusters of different sizes. Only at the upper boundary is it important not only that the innovator find the early adopters but that the early adopters be in a position to exert their collective influence on the early and late majorities. And in the threshold model, crossing the chasm is a dramatic development indeed, because any cascade that succeeds in toppling the vulnerable cluster will necessarily spread to the *entire network,* triggering a cascade of universal proportions. In the language of physics, the phase transition at the upper boundary is a *discontinuous phase transition* because the typical size of successful cascades jumps instantaneously from zero (that is, no cascades at all) to the entire system.

Cascades at the upper boundary of the cascade window are therefore even rarer and much larger than those at the lower boundary, resulting in a qualitatively different kind of unpredictability. Most innovations that occur in networks near the upper boundary die out before spreading very far at all, suppressed by the local stability of the individual nodes. This state of affairs can continue almost indefinitely, leading an observer to conclude that the system is, in fact, stable. And then, out of the blue, a single influence that initially seems no different from any other can overwhelm the entire network. There needn't be anything auspicious about the particular innovator who triggers such a cascade either. Unlike at the lower boundary, where connectors play a significant role in linking up vulnerable clusters, at the upper boundary connectivity is not the problem. Hence, cascades are almost as likely to be triggered by an individual with an average number of neighbors as someone to whom many people pay attention. When cascade propagation is dominated more by local stability than connectivity, being simply well connected is less important than being connected to individuals who can be influenced easily.

These features of the cascade window suggest some unexpected lessons for the diffusion of innovations, perhaps the most surprising of which is that a successful cascade has far less to do with the actual characteristics of the innovation, or even the innovator, than we tend to think. In the context of the cascade model at least, there is nothing that distinguishes the particular shock that triggers a global cascade from any other shock. Rather, all the action is generated by the connectivity of the vulnerable cluster to which the initial innovator is connected. And what makes the problem of determining success even more intractable is that the percolating vulnerable cluster, when it even exists, is a global property of the system, an elusive filament entwined throughout the entire network. It matters not only that a particular individual has one or more vulnerable neighbors, but also that those neighbors have one or more vulnerable neighbors, and so on. So even if you can identify potential early adopters, unless you can also view the network, you won't know whether or not they are all connected.

All this is not to say that factors like quality, price, and presentation are unimportant. By altering the adoption thresholds of individuals in the population, the innate properties of an innovation can still affect its success or failure. The point is that because thresholds do not alone determine the outcome, quality, price, and presentation can't either. In the regions of Figure 8.5, above and to the right of the cascade window (the point P, for example), the system can be altered to make it susceptible to global cascades either by lowering the average adoption threshold (the left arrow) or by lessening the connectivity of the network (the down arrow). In other words, the structure of the network can have as great an influence on the success or failure of an innovation as the inherent appeal of the innovation itself. And even inside the cascade window much of an innovation's fate hangs on random chance. If it hits the percolating cluster, it will succeed, and if it doesn't, it won't. As much as we may want to believe that it is the innate quality of an idea or product that determines its subsequent performance, or even the way it is presented, the model suggests that for any wild success, one could always find many equally deserving attempts that failed to receive more than a tiny fraction of the attention. It could just be that some innovations—*Harry Potter,* Razor Scooters, *Blair Witch Project*—hit just the right vulnerable cluster, while most do not. And in general no one will know which one is which until all the action is over.

A N O N L I N E A R   V I E W   O F   H I S T O R Y

THE NOTION THAT OUTCOMES CAN ONLY BE PROPERLY UNDERSTOOD in terms of the interactions of individuals, each of whom is reacting in real time to the decisions and actions of others, presents us with a quite different view of cause and effect than the one to which we are generally accustomed. Conventionally, when something or someone is success-ful, we assume that the extent of the success is proportional to some underlying measure of merit or significance. Successful artists are cre-ative geniuses, successful leaders are visionaries, and successful prod-ucts are just what consumers were looking for. Success, however, is a descriptor that can only be applied after the fact, and with hindsight it is easy to be wise. Our typically outcome-oriented view of the world, therefore, leads us to attribute the success of something to whatever characteristics it happens to exhibit, whether or not those characteris-tics were ever recognized as special beforehand.

What we don't generally consider is that the very same thing, with the very same characteristics, just as easily could have been a dismal failure. Nor do we typically waste much time lamenting the multitude of unsuccessful innovations that also could have been contenders had their circumstances been perhaps slightly different. History, in other words, has a tendency to ignore the things that *might* have happened but did not. Obviously, what actually happened is more relevant to our current circumstances than what didn't. But we have an additional predisposition to assume that the actual outcome was somehow *preferred* over all other possibilities, and this is where our perceptions of the world can misconstrue arbitrariness for order. From a scientific point of view, therefore, if we want to understand what might happen in the future, it is critical to consider not only what did happen but also what *could have* happened.

That accident and circumstance play important roles in history is hardly a new idea, but the notion of an information cascade suggests something more striking: that inputs and outcomes cannot be associated in a proportional, or even unique way. If a billion people believe in a particular religion, then we assume that the original message must have been inspired, otherwise why would a billion people believe in it? If a work of art is so much more famous than any other, it must surely be so much better, or else why would everyone talk about it? If a nation rallies around a leader to achieve great things, the leader must surely be great, or why would everyone follow him? So while greatness (or inspiration or fame) is always, in practice, conferred after the fact, our *perception* is that it was there all along—a necessary quality, intrinsic to the source of great change.

Before the fact, however, it is rarely clear what outcome any particular state of affairs will produce. And not simply because greatness, like genius, is hard to judge, or often misunderstood, but because it is almost never solely an intrinsic property in the first place. Rather, it is a consensus arrived at by large numbers of individuals, each observing the opinions of others as much as exercising their own independent judgment. People may believe simply because other people believe, people may talk about something simply because other people talk about it, and people may rally simply because other people are rallying. Such *contingent decision making* comprises the essence of an information cascade, and in so doing renders the relationship between initial cause and ultimate effect deeply ambiguous.

Psychologically, this view can be hard to accept—every era needs its icons, just as every revolution needs its leaders. But our tendency to credit innovators with an influence proportional to the ultimate outcome ignores the mechanism by which their actual influence was transmogrified into a mass movement. Just as with the stock market, when a major event appears in the historical record, we try to find what it was that preceded it, and when we find something—even if it was relatively minor in absolute terms—we attach to it great significance. According to Isaiah Berlin, Tolstoy's resentment of recorded history, especially military history, stemmed from his insight that amid the fog of war, no one—and especially not the generals—has any idea what is going on, and the balance between victor and vanquished tips more on the fulcrum of luck than under the forces of leadership or strategy. Yet after the smoke has cleared and the winner is revealed, it is the (accidentally) victorious general who gets all the glory.

From this perspective, Tolstoy probably would not have been any happier with late-twentieth-century science than he was with early-nineteenth-century warfare. Ever since the Celera corporation, led by J. Craig Venter, and the publicly funded consortium, headed by Francis Collins and Erik Lander, declared a tie in their race to sequence the human genome, Venter, Collins, and Lander have been sparring over who deserves credit for the breakthrough. In reality, none of them does: the genome project was a collaboration of hundreds, if not thousands, of hard-working scientists, without whom there would be no credit to disburse. In architecture, the situation is much the same. Frank Lloyd Wright, Eero Saarinen, and Frank Gehry are all revered for their striking designs, but without the teams of talented engineers and legions of construction workers who enable their drawings to actually stand up, none of these architects would have ever "created" a thing. The monumental maybe is too hard to comprehend directly, and so our minds react by representing an entire enterprise or period of history with a single person or part—an icon. Iconification, therefore, is an understandable cognitive device (and to be fair, many of our icons really are remarkably talented individuals), but it can mislead our intuition when we try to understand the origins of collective, as opposed to individual, behavior.

To take a more prosaic example, in early 1999, when Shawn Fanning was a nineteen-year-old student at Northeastern University, he designed a piece of code to help a friend download MP3 music files from the Internet. The result, a program they nicknamed Napster, became an overnight phenomenon, attracting tens of millions of users and the ire of the entire recording industry, and throwing Fanning into the midst of a worldwide commercial, legal, and ethical maelstrom. At least for a while, Fanning was the man in the middle, lionized by some and demonized by others, quoted in business papers and pictured on magazine covers. Before finally being forced to charge fees for its music-sharing services, Napster (now largely defunct) and Fanning had succeeded in striking a deal with the global publishing giant Bertelsmann. Not a bad effort for a college kid! Apparently not, but whose effort was it really?

The software that Fanning created was a neat trick, no doubt about it. But its enormous impact was a result not of any particular ingenuity of the code itself, or even of any particular vision of Fanning's—he was simply helping out a single friend. Rather, the magnitude of Napster's influence was a result of the huge numbers of people who realized that this was exactly what they were looking for, and who started to use it. Fanning didn't anticipate the unprecedented demand for his invention—he couldn't have. Probably not even the millions of eventual Napster users knew that they wanted to download free music off the Internet until suddenly the possibility presented itself, so how could Fanning? Actually, he didn't need to. All he had to do was release his idea, and once it was in the wild, a few people picked it up and started using it, leading a few more people to hear about it and to start using it as well. The more people used Napster, the more songs became available, hence the more attractive and visible it became to other people still.

If no one other than Fanning and a few of his friends had started using Napster, or if they hadn't had very good music collections or known enough other people who did, Napster may never have seen the light of day. To some extent, Napster had to be the way it was in order to succeed. If it had been expensive to download, impossible to use, or designed to do something for which there existed very little demand—like solve differential equations or translate Polish into Italian—it would never have become so popular. In terms of the threshold model, the adoption threshold had to be low enough for Napster to spread. But to some extent also, and possibly a very large extent, Napster's success was independent of its specific form and origin. And although Fanning, as its inventor, received most of the attention, the real engine that powered Napster from a mere idea to a phenomenon was the people who used it.

POWER TO THE PEOPLE

THE INNOVATORS AND REVOLUTIONARIES, IN OTHER WORDS, WHO act out of conscience, ideology, inventiveness, and passion are an essential component of a global cascade, forming the seed or trigger from which the cascade can propagate. But—and this is what makes cascades so hard to understand—*the seed alone is not enough.* In fact, as far as the success or failure of a cascade is concerned, seeds of change, much like their biological counterparts, are a dime a dozen. The seed that falls to earth may contain the blueprint of a fully flowered tree, and therefore in principle bears ultimate responsibility for the finished product. But its realization depends almost entirely on the nurturing qualities of the substrate on which it lands. Trees spread their seeds in profligate numbers for a reason: only one in very many will grow to fruition, and not because that one bears some special, unique quality, but because it *lands in the right place.* So it is for social seeds as well: the innovators and the agitators are *always* there, always trying to start something new and remake the world in their image. What makes their success difficult to predict is that in many cases, it has less to do with their particular vision and their individual characteristics than with the pattern of interactions into the midst of which their pinprick falls.

Like most generalizations, this statement is not always true. Sometimes individuals exert so profound an effect that their influence truly seems to have been guaranteed. When Einstein's original paper on special relativity was published in 1905, it overturned the scientific order of the past three hundred years, and from that moment on, Einstein's greatness was assured. Descartes and Newton also single-handedly revolutionized the scientific worldviews of their times— Descartes with analytical geometry and Newton with his universal theory of gravitation. Sometimes, in other words, a profound outcome implicates an equally profound cause. Breakthroughs of this nature, however, are exceedingly rare, and most social and scientific change is not wrought by giant cognitive leaps of singular genius. If one wants to trigger an avalanche in the mountains, one *could* drop an atomic bomb, but it's hardly necessary, and avalanches typically don't start that way. Rather, a single skier crunching through the wrong kind of snow on just the wrong part of the wrong mountain at the wrong time of day can unleash a fury that is grandly disproportionate to the cause.

And so it is apparently with cultural fads, technological innovations, political revolutions, cascading crises, stock market crashes, and other manners of collective madness, mania, and mass action. The trick is to focus *not* on the stimulus itself but on the structure of the network that the stimulus hits. In this respect, there is still a great deal of work to be done. Random networks, remember, are not very good representations of real networks, and work is currently under way to generalize the simplest cascade model to more realistic networks in which group structure, individual social identity, and mass media effects are included. The threshold rule also is a highly idealized representation of social decision making, and will require a number of embellishments if it is to be applied to any practical matter. But even now, some general insights are possible.

Perhaps the most striking property of the cascade model is that initial conditions that are indistinguishable before the fact can have dramatically different outcomes depending on the structure of the network. Quality (which here can be interpreted as the adoption threshold), therefore, is an unreliable predictor of success, and even great success is not necessarily a signature of great quality. The difference between a hugely successful innovation and an abject failure can be generated entirely through the dynamics of interactions between players who might have had nothing to do with its introduction. This is not to say that quality doesn't matter—it does, as do personalities and presentation. But in a world where individuals make decisions based not only on their own judgments but also on the judgments of others, quality is not enough.

ROBUSTNESS REVISITED

IN ADDITION TO ITS IMPLICATIONS FOR PREDICTABILITY, AN UNDER-standing of global cascades in networked systems can also shed light on the question of network robustness that we encountered in chapter 6. And in this context, we need not be discussing *social* contagion at all. Sometimes, systems that are characterized by many interdependent parts interacting in complicated ways, such as power grids and large organizations, can exhibit sudden, large failures despite all precautions taken to prevent them. The Yale sociologist Charles Perrow, who studied a series of organizational disasters from the partial meltdown at Three Mile Island to the *Challenger* explosion, calls such events *normal accidents*. Accidents, he argues, do not occur so much because of exceptional errors or inexcusable negligence but because a number of quite regular errors pile up, often compounded in an unanticipated way by the very same routines, reporting procedures, and responses that ordinarily keep things running smoothly. However exceptional they may appear, such accidents are best understood as the unexpected consequences of normal behavior; hence they are not only normal, but inevitable.

Perrow's position, outlined in his book *Normal Accidents,* might seem a bit pessimistic, but it closely resembles the picture of chronic unpredictability inherent in the cascade model. And this resemblance is more than metaphorical. Although we derived the threshold rule from the properties of social decision making, thresholds can arise in other contexts as well. Whenever the *state* of a node in a network can be represented as a choice between two alternatives—infected or susceptible, active or inactive, functioning or failed—that depends on the states of its neighbors, the problem is essentially one of contagion. And whenever the contagion exhibits dependencies between neighboring states, in the sense that the effect of one influence (like a failure) is compounded or alleviated by another, then a threshold rule arises. Hence, the cascade model can be applied not just to cascades of social decisions but also to cascades of failures in organizational networks and even power grids. As a result, the primary feature of the cascade model— that apparently stable systems can suddenly exhibit a very large cascade—can also be interpreted as a statement about the inherent fragility of complex systems, even those that seem robust.

A few years ago, John Doyle, a mathematician at the California Institute of Technology, and Jean Carlson, a physicist at the University of California at Santa Barbara, proposed a theory of what they call *highly optimized tolerance* (HOT) to explain the observed size distributions of a wide range of phenomena from forest fires to power outages. Their most striking conclusion was that real-world complex systems are invariably both robust *and* fragile. Because they have to survive in the real world, complex systems typically are able to withstand all manner of shocks, because either they were designed to or they evolved that way. If they could not, in fact, they would have to be modified or would cease to exist. But just as with the cascade model from before, every complex system has a weak point, which if struck in just the right manner, can bring down even the most painstakingly engineered house of cards. Once one of these weaknesses manifests itself, we usually rush to fix it, thus improving the system's robustness in some specific way (natural selection takes care of weaknesses in its own fashion). But as Doyle and Carlson demonstrate, that doesn't remove the fundamental fragility of the system—it is merely displaced until another day and possibly another kind of accident altogether.

Airplanes are a good example of this robust yet fragile phenomenon. Typically, as soon as a design flaw shows up in a major aircraft, sometimes by causing it to fall out of the sky, investigators pin down the origin of that specific problem. Every single plane of that type in the world is then checked and, if necessary, modified to prevent any repetitions of the problem. By and large, this is an effective procedure, as evidenced by the relative scarcity of recurring flaws leading to plane crashes. But it can't prevent plane crashes completely, for the simple reason that even the best maintenance procedures in the world can't be guaranteed to prevent faults that aren't yet known to exist.

And aircraft are children's toys compared with vast organizational machines like Enron and Kmart, both of which suddenly and unexpectedly declared bankruptcies in the month between December 2001 and January 2002 when I was finishing this chapter. In the real world, therefore, no amount of careful planning or even sophisticated science may be able to prevent disasters from happening occasionally. Should we just give up? Of course not, and neither Perrow, nor Doyle and Carlson, suggest that matters are hopeless. Rather, what is necessary is a richer conception of robustness. Not only should we design systems to avoid failures as much as possible, but we must also accept that failures will happen despite our best efforts and that a truly robust system is one that can survive even when disaster strikes. It is this conceptualization of robustness as a dual feature of a complex organization—preventing failures with one hand and preparing for them with the other—that we will explore in the next chapter.

CHAPTER NINE

Innovation, Adaptation, and Recovery

● —— ● —— ● —— ● —— ● —— ●

In JANUARY 1999, WHEN I WAS A POSTDOC AT THE SANTA FE Institute, I was giving a talk to representatives from the institute's business network, a group of companies that support the institute financially. Also present was Chuck Sabel, a law professor at Columbia whom I had met once or twice but knew mainly from his somewhat cantankerous reputation. I had given one too many talks on the small-world problem by that stage, and as I motored through my usual spiel, I was mostly hoping not to put anyone to sleep. So it was much to my surprise that as I was packing up afterward, Chuck came rushing up to me waving his hands urgently and insisting that we had to talk. Inasmuch as I understood anything about Chuck's work, it concerned the evolution of modern manufacturing and business processes, and so had nothing to do with me. Furthermore, I couldn't understand a word he was saying. Although Chuck, as I eventually discovered, is a fantastically interesting thinker, his manner is that of the intense, Harvard-trained intellectual that he is, replete with intimidating vocabulary, labyrinthine reasoning, and abstract conclusions. Listening to Chuck think is like drinking wine from a fire hose—it's good stuff, but it can still drown you.

After a few minutes of this, and with my eyes starting to water, I escaped by handing him the manuscript of a book I was working on, expecting that to be the last I'd ever hear from him. That was before I knew Chuck. A few days later, the phone rang. It was Chuck, and now he was *really* excited. Not only had he read the whole laborious manuscript (on the plane), but he was now convinced that his earlier instincts were correct and that the two of us should get together later that year to collaborate on a project. I still had no idea what he was talking about, but I was so taken by his enthusiasm that I agreed. By the time August rolled around, however, and Chuck showed up at Santa Fe, I was starting to panic. How was I supposed to spend an entire month working with a guy I hardly knew, on a project I barely understood? I was just beginning to write the whole situation off as an unfortunate mistake, when Chuck told me a story that has fascinated me ever since.

THE TOYOTA-AISIN CRISIS

In THE 1980s, JAPANESE AUTOMOBILE MANUFACTURING WAS THE envy of the world. Having mastered a suite of production processes like *just-in-time* inventory systems, *simultaneous engineering* (in which the design specifications of interdependent components are worked out concurrently rather than consecutively), and *mutual monitoring*, Japanese firms like Toyota and Honda had come to epitomize the concept of a modern lean corporation. Toyota, in particular, was held up to the world by management experts as a shining example of brutal efficiency happily cohabitating with creative flexibility. Churning out many of the world's best-engineered cars at prices that made even their European competitors wince, year after year Toyota was making Detroit look like an eight-hundred-pound gorilla trying to do aerobics.

It might come as a surprise then that the industrial behemoth that produces Toyota cars and trucks is much more than a single company. In actuality, it is a group of roughly two hundred companies integrated by their common interest in supplying the Toyota company itself with everything from electronic components to seat covers, and also by what is known as the *Toyota Production System*. TPS is a collection of the same kinds of manufacturing and design protocols that have been adopted by most Japanese (and these days American) industrial firms, so in a way it's nothing special. What makes it unique is the almost religious zeal with which it is implemented inside the Toyota group. Companies in the group, even those companies that compete with each other for Toyota's business, cooperate to an extent that almost seems counter to their interests. They routinely exchange personnel, share intellectual property, and assist each other at the cost of their own time and resources, all without the requirement of formal contracts or detailed record keeping. In many ways, they behave more like siblings than firms, striving for the approval of a watchful mother who cares at least as much that everyone get along as she does about performance.

This might seem like a good way to run a family, but it's not at all clear that it's a good way to build cars. Nevertheless, starting in the 1980s, U.S. firms, from automobile makers to microprocessor, software, and computer manufacturers, began to embrace Japanese production methods and conventions. Industry after industry was swept by Japanese-inspired trends, with *reengineering, total quality management,* and just-in-time inventory systems all taking their turn as flavor of the month. The end result of this upheaval was that American companies in the late 1990s looked very little like the vertically integrated hierarchies that had built cars for the likes of Henry Ford and Alfred Sloan in the 1920s and that had been the paradigmatic form of corporate order ever since. As hard as they have tried to change, however, the American auto giants have never quite been able to match the performance of their Japanese counterparts. And then, a few years ago, Toyota suffered a major crisis that left the global automotive industry with its collective mouth agape. Not only had Toyota's revolutionary production system finally gotten the company into terrible trouble, but just as quickly it had gotten them out.

In all of the Toyota group, one of the most significant and trusted members is a company called Aisin Seiki. Originally a division of Toyota itself, Aisin was spun off as a separate company in 1949 to concentrate on the specific business of manufacturing brake components. In particular, Aisin produces a class of devices called P-valves, which are used in all Toyota vehicles to help prevent skidding, by controlling pressure on the rear brakes. About the size of a pack of cigarettes, P-valves are not all that complicated, but because their role is so critical to safety, they must be manufactured precisely and so are produced in highly specialized facilities using custom-designed drills and gauges. On account of its spotless performance record, by 1997 Aisin had become Toyota's sole provider of P-valves. And for reasons of efficiency, Aisin had chosen to locate all its P-valve manufacturing lines in a single factory, the Kariya plant number 1, which at that time was producing 32,500 valves a day. Finally, because of the success of their just-in-time system, Toyota was holding only about two days' worth of P-valves in stock. Production at the Kariya plant was, therefore, an absolutely critical element of Toyota's supply chain. No factory, no P-valves. No P-valves, no brakes. No brakes, no cars.

Well, in the early morning of Saturday, February 1, 1997, the Kariya plant burned down. Just like that. By 9 A.M. that day, all the production lines for P-valves, along with those for clutch and tandem master cylinders, and most of the special-purpose tools that Aisin used for manufacture and quality control, had been destroyed. In just under five hours, Aisin's production capacity for P-valves had vanished almost entirely, and would take months to rebuild. *Months!* At the time, Toyota was rolling more than fifteen thousand cars a day off about thirty production lines. But by Wednesday, February 5, all production had ceased, rendering idle not only Toyota's own plants but also the facilities and workers of many of the firms whose business it was to supply them. Across the entire Kobe industrial zone, mighty factories lay silent, as the invincible Toyota group—like a broken Goliath—tumbled before the impact of a single well-placed stone.

Make no mistake, this was a first-class catastrophe, next to which even the giant Kobe earthquake two years earlier, paled in comparison.

What happened next, however, was every bit as dramatic as the disaster itself. In an astonishing coordinated response by over two hundred firms, and with very little direct oversight by either Aisin or Toyota, production of more than one hundred kinds of P-valves was reestablished within three days after the fire. As soon as Thursday, February 6, two of Toyota's plants had reopened, and by the following Monday, little more than a week after the crisis had begun, production of almost fourteen thousand cars a day had been restored. A week after that, the daily volume was right back at its predisaster level. Even so, the Ministry of International Trade and Industry estimated that the loss in output as a result of the fire amounted to one-twelfth of the entire Japanese transportation industry for the month of February.

At such a staggering rate of loss, the consequences of a months-long or even weeks-long shutdown would have been unthinkable. So clearly, whether simply to restore their own production or to curry future business with Toyota, every company in the Toyota group had ample incentive to cooperate. But as Toshihiro Nishiguchi and Alexandre Beaudet point out in their detailed account of the recovery effort, even strong incentives are not enough. No matter how much any of the individual firms in the Toyota group might have wanted to help, they still required the *capability* to do so. Bear in mind that very few of the sixty-two firms that became emergency producers of P-valves, or the more than 150 firms involved indirectly as suppliers, had any prior experience making the valves, nor did they have access to the kind of specialized tools that had been destroyed in the fire. One firm involved in the recovery effort, Brother Industries, was a sewing machine manufacturer that had never even made car parts! The interesting question, therefore, is not *why* did they stage so dramatic a recovery, but *how*?

Before the fires had even stopped burning, the Aisin engineers were on the job, assessing the damage and determining exactly what they needed done. They realized immediately that the recovery task, if it was to be performed fast enough to avert their now imminent doom, lay well beyond their capabilities as an individual firm and beyond the capabilities of their immediate suppliers. A much broader effort would be required, and one over which they would have little direct control. Later that same morning, having set up an emergency response head-quarters, Aisin sent out the distress call, defining the problem in the broadest possible terms and asking for help. And like fighters on the tarmac, scrambling at the siren, the companies of the Toyota group responded.

In this particular scenario, however, being helpful wasn't easy. Because the firms involved in the recovery effort lacked the tools and expertise specific to P-valve production, they were forced to invent on the fly, novel manufacturing procedures, solving both the design and production problems simultaneously. To make matters worse, Aisin's expertise rested largely with its own processes, and therefore it was often of little help in overcoming technical obstacles. And finally, in the whirlwind of the crisis, Aisin became extremely difficult to contact. Even after installing thousands of additional phone lines, so much information was flowing in and out in the form of queries, suggestions, solutions, and new problems, that the company was often unreachable, leaving the cavalry largely to its own devices.

This, however, is where all the training kicked in. After years of experience with the Toyota Production System, all the companies involved possessed a common understanding of how problems should be approached and solved. To them, simultaneous design and engi-neering was an everyday activity, and because Aisin knew this, they were able to specify their requirements to a minimum level of detail, allowing potential suppliers the greatest possible latitude in deciding how to proceed. Even more important, while the particular situation was unfamiliar, the idea of cooperating was not. Because many of the firms involved in the recovery effort had previously exchanged person-nel and technical information with Aisin, and also with each other, they could make use of lines of communication, information resources, and social ties that were already established. They understood and trusted each other, an arrangement that facilitated not only the rapid dissemination of information (including even descriptions of their mistakes) but also the mobilization and commitment of resources.

Some firms went as far as to completely rearrange their production priorities in order to assist in the effort, choosing either to cut back on other jobs or outsource the less technically demanding ones to their suppliers. Other firms commandeered drills and gauges from all over the country, out of shop windows, and even from the United States, without worrying about how all this disruption would be paid for. In effect then, the companies of the Toyota group managed to pull off two recovery efforts simultaneously. First, they redistributed the stress of a major failure from one firm to hundreds of firms, thus minimizing the damage to any one member of the group. And second, they recombined the resources of those same firms in multiple distinct and original configurations in order to produce an equivalent output of P-valves. They did all this without generating any additional breakdowns, with very little central direction, and almost completely in the absence of formal contracts. And they did it in just three days.

Thanks to researchers like Nishiguchi and Beaudet, we have a reasonable record of the events that led up to and followed the Aisin crisis. So in some sense we know how it was resolved, and what it was about the companies in the Toyota group that enabled them to solve it. But in the same way that the mere sequence of failures in the power transmission grid couldn't tell us why the system was vulnerable to a cascade in the first place, or that a mere historical telling of a cultural fad doesn't reveal why an entire population of individuals should all of a sudden prefer one thing over another, the account itself can't tell us what enabled the system to survive such a massive shock.

As with the power grid example, the failure of a single component in a very large system turned out to have global repercussions, generating a widespread catastrophic failure. But the Aisin case is different in that the system subsequently recovered almost as rapidly as it had succumbed, and with little centralized control. It would be as if the power grid, having suffered the very same cascading failure that sent it reeling in August 1996, had gotten back on its feet in a matter of a few hours while the controllers sat there wondering what was going on. Such "self-healing" systems are as yet little more than a gleam in the engineer's eye, yet in the world of organizations it seems they have already been built. So what can we learn from the Aisin crisis that can help us understand the design of systems that can recover even from potentially devastating failures? And more generally, what can the Toyota group teach us about the architecture of modern industrial organization? How is it, in other words, that the performance of firms—meaning their ability to allocate resources, innovate, adapt, and solve problems, both routine and radical—is related to their organizational architecture?

MARKETS AND HIERARCHIES

INDUSTRIAL ORGANIZATION IS ACTUALLY QUITE AN OLD SUBJECT, growing out of the economic and social upheaval of the Industrial Revolution. Indeed, it is the topic of industrial organization on which Adam Smith opens his monumental treatise, *The Wealth of Nations*. In particular, Smith discusses the *division of labor*, the principle that he originally inferred from his observations of workers in manufacturing plants, who consistently performed better when their collective task was broken down into specialized subtasks. The example he uses to illustrate the principle is, of all things, the production of pins. Although it seems trivial, making a pin involves over twenty independent steps, like extruding the wire, grinding the point, hammering the head, cutting the wire, and so on. At the time of Smith's writing, in the late eighteenth century, even a skilled artisan, when working alone, could only make a handful of pins a day. Yet Smith observed that when the labor was divided among a team of ten men, each performing only one or two of the steps and using specialized tools, literally thousands of times that amount could be produced.

That teams of workers performing specialized elements of a complex task can produce much more than the same number of workers when each is performing the task in its entirety is actually a fundamental consequence of human learning. A very general rule, called *learning by doing,* holds that the more frequently we do something, the better we get at it. And if we have fewer tasks to do, and therefore can do each one more often, it follows that when we perform only a single step of a production process, we perform it more efficiently than if we have to do all the other steps as well. The benefit that each worker derives from learning to do one task efficiently is called *returns to specialization.* And by allocating the components of a complex process to many different individuals, working in parallel, the division of labor harnesses returns to specialization many times over.

According to the division of labor, therefore, the more specialized an individual's job, the better. In building an automobile, for example, the obvious subtasks would correspond to the major components of the vehicle—the body, engine, transmission, interior, and so on. Any one of these components, however, is a complex task in itself, thus further layers of specialization are required. The engine, for example, can be divided into the engine block, the fuel supply, the cooling system, and the electrical system, each of which, in turn, may require further decomposition still, and so on, until the entire complex task has been *partitioned* into a set of elementary steps. And because each of these steps generates returns to specialization, the overall gains in efficiency are vast.

So profound were the returns to specialization that Smith observed, he proposed the division of labor as the fundamental distinguishing feature of civilized societies. In societies without specialized labor, every family has to supply all its own needs, including food, clothing, shelter, and all the artifacts of everyday life. In such a world, merely surviving is a full-time job, and each generation is compelled to start from essentially the same step as the last. Schools, governments, and professional armies can't exist, nor can manufacturing, construction, transportation, or service industries. But as central as the division of labor was to Smith's view of industrial organization, he never actually specified the mechanism by which the specialized subtasks should be aggregated into a complex whole. In *The Wealth of Nations,* Smith skirts the issue, claiming only that the extent to which specialization is possible depends on the "extent of the market." By this statement he meant that the larger the pool of potential consumers, the more resources a firm can afford to invest in building production facilities, designing and creating specialized machinery, and employing workers, thereby benefiting from economies of scale. But this description doesn't specify why it should be formal entities called "firms" that are responsible for production, instead of, say, independent contractors, temporary workers, or consultants.

Nor does the division of labor necessarily imply that firms, when they do exist, should resemble the hierarchical authority structures that form our image of nineteenth- and early-twentieth-century industrialization. Just because tasks can be accomplished more effectively by breaking them down, in a hierarchical fashion, into more and more specialized subcomponents, doesn't imply *on its own* that firms have to be organized in the same way. Nevertheless, because many firms after the Industrial Revolution *were* actually organized in just this way, the consensus of economic theory for much of the last century has been that the optimal form of industrial organization, and, by association, the internal architecture of a business firm, is a hierarchy.

To cut a (very) long story short, the most generally agreed-upon economic theory of industrial organization essentially divides the world between hierarchies and markets. Firms, it claims, exist because markets in the real world suffer from a set of imperfections that the Nobel Prize–winning economist Ronald Coase called *transaction costs.* If everyone could discover, draw up, and enforce market-based contracts with everyone else (if we could all be independent contractors, for example), then the immense flexibility of market forces would effectively eliminate the need for firms entirely. But in the real world, as we have already seen in a number of contexts, information is costly to discover and hard to process. Furthermore, any agreement between two parties, even if it seems like a good idea at the time, is subject to uncertainty about future conditions and unexpected eventualities. If a contract to which both parties agreed at one point in time suddenly seems like a bad idea to one of them, that party may decide to renege, probably to the loss of the other party. And contract enforcement is difficult and expensive in a world where ambiguity and unanticipated circumstances can cloud the clearest of intentions.

Coase's main claim, therefore, was that firms exist in order to sweep away all the costs associated with market transactions, replacing them with a single contract of employment. Inside a firm, in other words, markets cease to operate, and the skills, resources, and time of its employees are coordinated through a strict authority structure. Although Coase himself never specified what this authority structure should look like, the consensus of subsequent economic theory is that it should be a hierarchy. Markets, meanwhile, continue to operate between firms, where the boundary between firm and market is a trade-off between the *coordination cost* of conducting a particular function within the firm and the transaction cost of striking an external contract. If the relationship between two firms ever becomes so specialized that one is effectively in a position to manipulate the other, the problem is assumed to be resolved by a merger or an acquisition. Hence, firms grow by the process of *vertical integration:* one hierarchy effectively gets absorbed into another, generating a larger, *vertically integrated* hierarchy. Conversely, when a firm decides that some internal function is too expensive, it either spins off that branch of the hierarchy to form a specialized subsidiary, or eliminates it altogether, outsourcing the function to another firm. Whatever the scenario, firms remain hierarchies (only their size and number change), and markets operate between firms.

It really is an elegant theory, and has such a ring of plausibility that it has dominated economic thinking on firms for more than half a century. But in 1984, a revolutionary book written by two MIT professors, an economist and a political scientist, fired the first warning shots in what has become an increasingly tangled conflict over the true nature of industrial organization and the future of economic growth. The book was called *The Second Industrial Divide,* and the political scientist of the pair was Charles F. Sabel—the very same Chuck Sabel who accosted me in Santa Fe fifteen years later.

INDUSTRIAL DIVIDES

FROM AN ECONOMIST'S PERSPECTIVE, PERHAPS THE MOST polemical (if not the most significant) point that Chuck and his coauthor, Michael Piore, made is that the theory of the firm came about essentially *after the fact.* Only after large-scale industrialization had effectively settled on the model of vertical integration and its associated economies of scale did economists start to develop a theory of the firm. And as a consequence, it was only a *particular* type of firm—the large, vertically integrated hierarchy—that they tried to explain, as if no other theory of industrial organization could even make sense. But looking back at the late nineteenth century, when the modern image of the industrial firm first emerged, Piore and Sabel showed that the hierarchy was not the only successful form of industrial organization, nor was its eventual preeminence necessarily based on universal economic principles.

Vertical integration, of course, did not become the dominant form of industrial organization by accident—for a variety of reasons, it made perfect sense at the time. What Piore and Sabel claimed, however, was that organizational forms arise as the solution to problems that are partly economic and partly social, political, and historical. The strongest manifestation of the noneconomic dependence of economic decisions is that technological history occasionally encounters branch points, what they called *divides,* at which a decision is made between competing solutions to a general problem. And once a decision is made, the winning solution gets so locked into contemporary and historical thinking alike that the world forgets it ever had an alternative.

Piore and Sabel argued that the first such *industrial divide* was the Industrial Revolution itself. During this time, the vertically integrated model of huge factories, highly specialized production lines, and generally unskilled labor outcompeted and nearly eliminated the previously dominant *craft* system of highly skilled craftsmen operating general-purpose tools and machines. For nearly a century thereafter, industrial organization followed the hierarchical model. And like researchers focused on a particular scientific paradigm, economists, business leaders, and policy makers simply assumed that no other form of organization was conceivable. The division of labor, industrial organization, and vertical integration were all thought to be interchangeable concepts.

By the late 1970s, however, the world had begun to change. The rapid growth of the world's postwar industrialized economies had begun to reach the limits of what their domestic consumer markets could demand, and further growth required a dramatic globalization of both production and trade. Around the same time, and partly for the same reason, the fixed currency exchange rate system of the 1944 Bretton Woods agreement began to break down, and the first cracks appeared in the walls of trade protection behind which many nations' postwar reconstruction strategies had sheltered. Exacerbating these tectonic changes in the global economy were a series of economic and political shocks—two oil crises in quick succession, the Iranian revolution of 1979, and a combination of growing unemployment and inflation in the United States and Europe—all of which eroded the industrialized world's vision of an endlessly prosperous future. Within the span of a decade, the world had become a murkier, more uncertain place, and business leaders had to start thinking outside the box of conventional economic wisdom in order to survive. Although it was clear to anyone paying attention that the postwar prosperity party was over, no one seemed to recognize that the old economic order itself had been overturned—that the world was, in fact, entering its second industrial divide.

The *Second Industrial Divide* was therefore partly an economics version of the emperor's new clothes and partly an attempt to sketch out an alternative, better-clothed point of view. The craft system, Piore and Sabel pointed out, had never entirely gone away, having persisted in the manufacturing regions of northern Italy and even parts of France, Switzerland, and the United Kingdom. In part, it had survived in those places because of their unique histories, the social networks that existed between traditional family-based production systems, and the geographical concentrations of specialized skills they represented. But craft production had survived also on its merits, outperforming vertically integrated economies of scale in fast-moving and unpredictable industries like fabric production, which depends for its livelihood on the ever transient world of fashion.

Far more important than the persistence of the craft system itself, however, was that its essential feature, what Piore and Sabel dubbed *flexible specialization,* had slowly been adopted by a multitude of firms, even in the most stalwart economies-of-scale industries. The U.S. steel industry, for example, has spent the past thirty years abandoning its traditional blast-furnace technology in favor of smaller, more flexible mini-mills. Flexible specialization is the antithesis of a vertically integrated hierarchy in that it exploits *economies of scope* rather than economies of scale. Instead of sinking large amounts of capital into specialized production facilities that subsequently produce a restricted line of products quickly and cheaply, flexible specialization relies on general-purpose machinery and skilled workers to produce a wide range of products in small batches.

Returns to specialization, remember, derive from the frequent repetition of a limited range of tasks, and repetition is only possible if the tasks themselves don't change. In slowly changing environments, therefore, in which generic products appeal to large numbers of consumers and the range of competing choices is limited, economies of

scale are optimal. But in the rapidly globalizing world of the late twentieth century, with firms pinned between uncertain economic and
political forecasts on the one hand, and increasingly heterogeneous
tastes of consumers on the other, economies of scope gained a critical
advantage. Uncertainty, ambiguity, and rapid change, in other words,
favor flexibility and adaptability over sheer scale. And in the two
decades since Sabel and Piore first pointed this fact out, the world of
business has become only more and more ambiguous.

Recently I asked Chuck how he felt the ideas in his book were holding up almost twenty years after he had first espoused them. Had he
and Piore been proved right? Well, yes and no. Yes, in the sense that the
dominance of so-called new organizational forms over traditional vertically integrated hierarchies was now essentially unquestioned (except
perhaps in the more conservative economics journals). And yes, in the
sense that the reason for this shift was generally agreed to be the sharp
increase in uncertainty and change associated with the global business
environment of the past few decades, in old-economy staples like tex-
tiles, steel, automobiles, and retail as well as the new-economy industries of biotechnology and computing. But there was a sense in which,
over the past ten years in particular, Chuck had come to see their proposed solution of flexible specialization as critically incomplete.

AMBIGUITY

THE IDEA UNDERLYING FLEXIBLE SPECIALIZATION IS ROUGHLY
that the tasks required of a modern firm—whether building an automobile, creating a new weave of fabric for the spring catalogue, or
designing the next computer operating system—are subject to significant unpredictability and rapid change. Under those circumstances,
instead of sinking large amounts of capital into specialized production
facilities, a firm will adopt an economies-of-scope approach, fostering
flexibly organized teams of highly skilled workers that can repeatedly

and rapidly recombine their specialized skills to produce small batches of a wide range of goods. It sounds like a powerful formula, and it is. However, it glosses over a second kind of ambiguity, which is of a fundamentally deeper character. Not only do firms face uncertainty over which particular task is required of them by the external marketplace, but also they are uncertain about precisely how they should go about completing *any* task, or what the corresponding criteria for success might be.

At the bottom of the mystery, and implicit in almost all theories of the firm, is the assumption that even if the accomplishment of a complex task is a decentralized process, requiring the simultaneous, coordinated efforts of many specialized workers, its design is somehow centralized, imposed in a sense "from above." What Chuck had come to realize in the years since the *Second Industrial Divide* was that this assumption is a convenient fiction. In reality, when a firm embarks on a major new project, *the people involved don't actually know how they are going to do it*. In fast-moving industries from software to automobiles, designs are rarely final before production itself has commenced, and performance benchmarks evolve along with the project. Furthermore, no one person's role in the overall scheme is ever precisely specified in advance. Rather, each person starts with a general notion of what is required of him or her, and refines that notion only by interacting with other problem solvers (who, of course, are doing the same). The true ambiguity of modern business processes, in other words, is not just that the environment necessitates continual redesign of the production process but that *design itself*, along with innovation and trouble shooting, is a task to be performed, not only at the same time as the task of production but also in the same decentralized fashion.

When *environmental ambiguity* is low—that is, when change occurs slowly and the future is predictable—then this fundamental *task ambiguity* is suppressed, effectively allowing the design/learning and production phases to be completed separately. In a sufficiently slowly changing world, individuals participating in even the most complex tasks have sufficient time to pass through their learning phase and settle into the business of routine production. The effect is that the division of labor among the individuals comprising a firm mirrors the hierarchical partitioning of the task itself, hence the persistent hierarchical image of firms.

But once the environment cranks up the rate of change required for competitive performance, complex tasks must be correspondingly repartitioned, and available human capital correspondingly reallocated. And absent some infinitely capable overseer, this repartitioning problem must be solved by the same individuals who have to perform the task of production. The result, in a successful firm, is a continual swirl of problem-solving activity and ever shifting interactions between the problem solvers, each of whom has information relevant to the solution of a particular problem but none of whom knows enough to act in isolation. Nor does any one person know precisely who knows what; hence, problem solving is a matter not just of forming the necessary combination of resources (this is what flexible specialization was about) but of searching for and discovering those resources in the first place.

This process is necessarily an inexact science, but it can be done. In Honda manufacturing plants, for example, even relatively routine manufacturing problems are solved by rapidly created, temporary teams drawn as needed from across the plant, not just the specific area in which the problem was first observed, and including line workers, engineers, and managers. The reason is that even straightforward-seeming problems can have far-reaching roots and, thus, can require a surprisingly broad range of institutional knowledge for their resolution. A simple paint defect, for example, observed in the final inspection stages of the production line may result from a faulty valve, which might have stopped working because a particular spray station is continually overtaxed, because another spray station is never working, because that spray station has a problem with its computer control mechanism, which resulted from an incorrect software setup, which can be traced to an overworked systems administrator, who is spending too much time helping managers with their e-mail accounts, and so on. No single person can know all this, but companies like Honda and Toyota have discovered that with a sufficiently diverse portfolio of participants, even quite complicated causal chains can be identified quickly.

It seemed to Chuck that routine problem-solving activity not only was a defining feature of modern firms responding to an ever more ambiguous business environment. Consequently, it was critical to understanding both the complicated structure of industrial organizations like the Toyota group and their ability to recover from large failures like the Aisin crisis. The formal theory of firms, however, had not kept pace with the changing phenomenon itself. Economists, although eager to build analytically rigorous models, had been unwilling to acknowledge the ambiguity inherent in modern industrial organization or to incorporate it in their theories. Hence, economic theory was essentially stuck in the era of the market-hierarchy dichotomy, a view that effectively ignored problem solving and failure altogether. Sociologists and business analysts, meanwhile, were far more comfortable with the notion of adaptability and robustness. But they had been unable to put the kind of analytical teeth into their models that would enable them to present a theoretically compelling alternative to the apparent optimality of markets and hierarchies. To Chuck it seemed clear that some other approach was required.

THE THIRD WAY

BY THE TIME I MET HIM, CHUCK HAD BECOME CONVINCED NOT only that ambiguity and problem solving were central to the behavior of firms, but also that with just the right mathematical framework, he could understand how. As he once said to me, "I know what the answer should look like, and if I were a mathematician I could just write it down. But I'm not." All of which explains his excitement that day in Santa Fe. In the hour of listening to me talk about small-world networks, he sensed that the models Steve and I had developed were capturing some of the features he felt were important. In firms, just as in social networks, individuals make decisions about whom to connect with, and although these decisions are based on the individuals' local perceptions of the network, they can have global consequences. In particular, Chuck was intrigued by the dramatic effects of random rewiring: individuals in tight-knit teams (clusters) engage in problem-solving searches that connect them to previously distant parts of the organization (random shortcuts), thus enhancing the coordination capability of the firm as a whole (reduced path length). The parallels between the two problems seemed striking, and we thought that a month would be long enough to sort out the subtle differences between them. As the weeks stretched into months, however, and the months into years, we finally conceded that the differences were both more important and more elusive than we had anticipated.

Eventually we decided that we needed some help. This was right around the time when I was moving back to New York to join the Sociology Department at Columbia, after a two-year sojourn in Santa Fe and Boston. As luck would have it, a friend of mine, a mathematician by the name of Peter Dodds (who appeared briefly back in chapter 5), was moving to New York also. And how Peter and I came to be in the same place at the same time is something of a small-world story in itself. A fellow Australian, Peter had migrated to the United States a year after I had, to study mathematics at MIT with none other than Steve Strogatz. Unfortunately for Peter, Steve was leaving the very next week to start his job at Cornell. I remember that when Steve and I first started working together, he did mention to me that another Australian had showed up at MIT just as he was leaving and had been very disappointed to see him go. But neither of us ever thought to hear about him again.

Two years later, I was at Thanksgiving dinner in Ithaca with some other orphaned Australians, and we began talking about the small-world research that I had just started. One of them, the brother of another Cornell student, was visiting from Harvard. After listening to me for a while, he mentioned that his friend Peter would be really interested in this stuff, and that he had come to MIT to work with this guy—Steve somebody—who had jumped ship to Cornell. "That's my adviser," I said, and there it rested again, until over two years later, at the Santa Fe Institute. One day my office mate, Geoffrey West, a distinguished physicist and expatriate Brit, mentioned that he was inviting "one of your fellow countrymen" from MIT to recruit him for a postdoc position. "Oh," I said, "let me guess . . . his name's Peter, right?" Sure enough. And that's when I finally met Peter. He didn't take the job, however, preferring to stay on at MIT to work with his Ph.D. adviser, Dan Rothman (who, you will not be surprised to learn, was a friend of Steve's). Dan, however, visited Santa Fe as well, soon after Peter, which is how I met Dan, which is how I got invited to give a talk at MIT some months later, which is how I met Andy Lo, which is how I got my job at MIT, which is how Peter and I finally became friends. A year later, we both landed jobs at Columbia and moved to New York within weeks of one another.

It seemed only proper that after years of being dragged into each other's orbits by our common interests, backgrounds, and even friends, we should do some work together, and so I told Peter about the problem I was working on with Chuck. Having done his dissertation on the branching structure of river networks, Peter knew the mathematical literature on networks. But his focus at the time was in earth science and biology, so he was a little reluctant to throw himself into the foreign world of sociology and economics. Once he met Chuck, however, and started to appreciate the magnitude of the problem, his curiosity quickly got the better of him, and he was soon fully on board. It still took us a while before we made any progress, but along the way, we started to see some connections between the specific problem we *thought* we were studying—the role of ambiguity and problem solving in firms—and the more general problem of robustness in networked systems like the Internet that need to survive and function in the face of unpredictable breakdowns and patterns of user demand.

COPING WITH AMBIGUITY

THE PROBLEM WITH AMBIGUITY, IT EVENTUALLY DAWNED ON THE three of us, is that it's . . . *ambiguous*. How on earth do you define ambiguity precisely, when its very nature—the reason why it is a problem to be dealt with in the first place—is that it is too slippery to pin down? Yet we did need to pin it down—otherwise it wouldn't be possible to specify how one organizational form could cope with it better than another. The trick, we decided, was to tackle ambiguity indirectly by focusing on its *effects* rather than its origins. When solving complex problems in ambiguous environments, individuals compensate for their limited knowledge of the interdependencies between their various tasks and for their uncertainty about the future by exchanging information—knowledge, advice, expertise, and resources—with other problem solvers within the same organization. Ambiguity, in other words, necessitates communication between individuals whose tasks are mutually dependent, in the sense that one possesses information or resources relevant to the other. And when the environment is rapidly changing, so too are the problems; hence, intense communication becomes an ongoing necessity.

The problem of coping with chronic environmental ambiguity is, therefore, equivalent to the problem of distributed communication. Firms that are bad at facilitating distributed communications are bad at solving problems, and therefore bad at handling uncertainty and change. Our strategy, therefore, was to think about the organizations as *networks of information processors*, where the role of the network was to handle large volumes of information efficiently and without overloading any *individual* processors. On the surface, this problem sounds a lot like the ones we have encountered earlier in this story. Whether considering the spread of a disease or a cultural norm, whether searching for a distant target or retaining connectivity in the face of failures, many network problems boil down to the transmission of information in connected systems.

A critical difference between organizational networks and the network models presented in earlier chapters, however, is that organizations exhibit an intrinsically hierarchical nature. The traditional view of the firm as a vertically integrated hierarchy may be critically incomplete, but that doesn't make it irrelevant. Although, as we will see, hierarchies respond poorly to ambiguity and failure, they are excellent structures for exerting control. And control remains a central feature of business firms and public bureaucracies alike. Individuals may report to multiple bosses, or report to different bosses at different times, but even in the most free flowing of new-economy firms, everyone still has a boss.

Hierarchies, furthermore, are not confined to the organization of individuals within the firm. Much of large-scale industrial organization, from industry groups like Toyota's to the structure of entire economies, is predicated on the notion of a hierarchy. Even many physical networks are designed on hierarchical principles (although, as we will see, they are typically not *pure* hierarchies either). The Internet, for example, comprises a backbone of large hubs, descending through layers of successively smaller providers to the ultimate level of individual users. The airline network is very similar. So as much as we need to get away from an exclusively hierarchical view of firms, the hierarchy is not just an endemic feature of modern business but an important one. That the vast majority of network models have either ignored hierarchies altogether or else ignored anything *but* a hierarchy leaves us once more in essentially uncharted terrain.

Another feature of organizational networks that distinguishes them from other kinds of networks we have discussed is that individuals are limited in how much work they can do. This constraint has serious implications for both the production and the information-processing tasks that modern organizations must perform. From a production perspective, efficiency requires that an organization constrain the nonproduction activities of its workers. One way to think about this is that network ties are costly, in terms of time and energy. And because individuals have finite amounts of both, it follows that the more relationships one actively maintains at work, the less actual production-related work one can do. Production efficiency, in fact, is the reason why hierarchies dominate the economics literature on firms. By adding more and more levels, through vertical integration, a purely hierarchical network can grow very large without any one individual having to supervise more than a fixed number of immediate subordinates, a constraint that in economics is called the *span of control*.

But individuals in problem-solving organizations must not only supervise subordinates but also coordinate their activities. In this (admittedly simplistic) view of the world, a true manager doesn't actually "do" anything in the traditional production-oriented view of industry. I used to wonder about this while sitting on the Delta Shuttle between Boston and New York and listening to all the executives and consultants on their cell phones frantically arranging some terribly important meeting. "What," I wondered, squeezed in between parallel displays of urgency, "do these guys actually produce?" If all a person does is rush from one meeting to another, what does he or she actually contribute to the productivity of the organization? The answer, from an information processing point of view, is that a manager's principal task is not production at all but *coordination,* to serve as an information pump between the individuals whose task *is* production. From this perspective, meetings are merely an institutionalized means of exchanging information between different parts of the organization, as are the annual conferences, task forces, and committees that often seem like a waste of time to an outside observer (and sometimes to the participants themselves). All pumps, however, including information pumps, have limited capacity. Even the most competent and energetic manager can only attend so many meetings, log so many frequent-flier miles, and field so many requests for information before being overwhelmed.

A *robust* information-processing network, therefore, is one that distributes not only the production load but also the burden of information *redistribution* as evenly as possible, thus maximizing the volume of information that can be processed without suffering breakdowns. And hierarchies, although they make highly efficient distribution networks, are extremely poor at redistribution. Imagine, for example, an organization in which every activity must be monitored, coordinated, and approved by a formal chain of command. In theory, such strictly hierarchical organizations do exist, the army being perhaps the quintessential example. But in practice, as soon as any ambiguity enters the picture, the chain of command is immediately saturated by the demands of processing endless requests for information and guidance. To see this, pick a *source* node (S) at random from the hierarchy in Figure 9.1 and imagine that it sends a message to another *target* node (T), possibly a request for information or assistance. In a pure hierarchy, the request must be

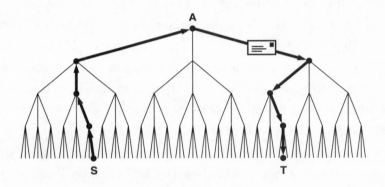

Figure 9.1. In a pure hierarchy, all messages between nodes must be processed by the chain of command, forcing nodes at the top to process information passed between many pairs of nodes lower down. Here, A is the lowest common ancestor of nodes S (the source of the request) and T (the target).

passed up the chain of command until it reaches the *lowest common ancestor* node (A), at which point it can be relayed back down to the target. The successful transmission of the request depends on every node in the chain performing its information-processing duty, but not every node is burdened equally. As Figure 9.1 implies, the higher up the chain of command a node sits, the more pairs of source and target nodes will pass messages through it, hence the greater its information-processing burden. In a pure hierarchy operating in an ambiguous environment, the burden of information processing is so unevenly distributed that unless something is done to accommodate it, the hierarchy will fail.

In physical information-processing networks like the Internet, the increasing burden associated with a higher position in the hierarchy can be compensated for (although not entirely) by increasing the capacity of the relevant servers and routers. The routers of the Internet backbone, for example, have much greater processing capacity than does the link from your computer to your local Internet service provider (ISP), or even the link from the ISP to the backbone. The reason is again implied by Figure 9.1: millions of individuals like you are attempting to send messages along the backbone, whereas far fewer users share your ISP (in Figure 9.1, the ISP for S would be its immediate superior in the hierarchy). But in organizational networks, one cannot simply increase the size and speed of one's brain just because there is more work to be done. Naturally, some people work harder or more efficiently than others, but it remains the case that unlike computers, people are simply not *scaleable*. So if the rate of problem-solving activity is to increase or even if the organization simply grows in size, then the resulting pressure on the chain of command must be relieved in some other way.

An obvious approach is to bypass the overtaxed node by creating a shortcut, thus rechanneling the congestion through an additional network tie. Building and maintaining new ties, however, leaves individuals less time for production; hence, both congestion and ties are costly. What is the most efficient way to balance these two kinds of costs? In our work on small-world networks, Steve and I found that the addition of a single shortcut contracted the paths between many distant pairs of nodes simultaneously, thereby effectively reducing congestion along many long chains of intermediaries. By dramatically reducing the average separation between nodes, by making the world small, random shortcuts seem like a powerful way to reduce congestion. But there are two big problems with this purely random approach. First, it doesn't account for the stratification by rank that characterizes a hierarchy. And second, by allowing each shortcut to contract the distance between many pairs of nodes simultaneously, it assumes that the shortcuts in question have an effectively unlimited capacity to transmit data. However, as already emphasized, individuals in organizations have capacity constraints, so any one link can relieve overall congestion by only so much. The result, as we can see from the top line in Figure 9.2, is that the random addition of links reduces the burden of the most congested nodes only slowly and is therefore of relatively little use in preventing failures. Just because the world is small does not necessarily make it either efficient or robust.

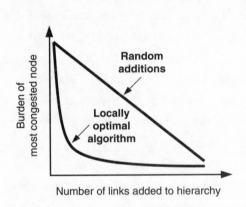

Figure 9.2. Adding bypass links to the hierarchy reduces the burden of the most congested nodes, but the effect can vary greatly depending on how the links are added. When they are added at random (top line), many links are required to generate a significant reduction. But when they are added according to the method shown in Figure 9.3, a few links can generate a large effect.

MULTISCALE NETWORKS

IF ADDING LINKS UNIFORMLY AT RANDOM ISN'T A GOOD WAY TO reduce information congestion, what is? In general, this is a hard question to answer, requiring as it does a balance between local capacity constraints and global (system-wide) performance. Fortunately, the stratified nature of the hierarchy yields a simple local strategy, illustrated in Figure 9.3, that is surprisingly close to optimal. Because all information processing in the model is generated by nodes passing messages to their immediate network neighbors, it follows that the burden of any particular node can be relieved by the greatest possible amount by connecting the two neighbors for whom it relays the most messages. That this purely local strategy should also work to reduce

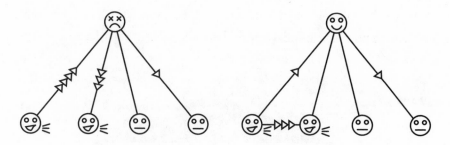

Figure 9.3. A locally optimal algorithm. The most congested node
is bypassed by adding a link between its two neighbors for whom it
relays the most messages (indicated by arrows).

global congestion by close to the optimal amount is not entirely obvious. After all, the messages are not eliminated, they are simply rerouted, so you might think that this would increase congestion elsewhere in the system. But because the strategy always selects the most congested node to relieve, and because (as Figure 9.3 indicates) the nodes that it connects were handling those messages anyway, the effect is always to reduce overall congestion without increasing any individual's burden.

As the bottom curve of Figure 9.2 indicates, this simple local algorithm for adding links appears to relieve information congestion effectively under a broad range of environmental conditions, far more so than a purely random approach. But precisely what kind of network *structure* it generates in the process depends a good deal on the *kind* of problem solving necessitated by the environment. When problem solving is purely local, requiring messages to be passed between members of the same work team, for example, or subscribers to the same ISP, congestion can be relieved effectively by a process that corresponds to *team building.* Supervisors who try to micromanage their subordinates typically find themselves overworked whenever the group collectively faces a difficult problem. Locally, the solution—allowing team members to work together without direct supervision—is more or less that described in Figure 9.3. The global picture that emerges as a consequence of these local changes, meanwhile, is represented schematically in Figure 9.4, in which *local teams* (coworkers who have the same immediate boss) form independently at all levels of the hierarchy.

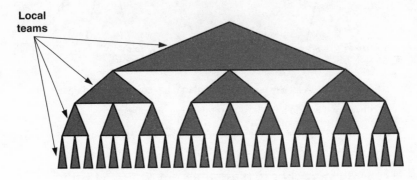

Figure 9.4. When message passing is purely local, the optimal network structure consists of local teams at all levels of the hierarchy.

At the other extreme, where message passing occurs exclusively between pairs of individuals that are distantly separated (in different divisions of a firm, say), almost all the burden of information processing is shifted to the top of the hierarchy. As Figure 9.5 indicates, the resulting network can be divided (approximately) into two tiers: a densely connected *core*, analogous to a headquarters or a central processing unit, and a *periphery* of pure distribution hierarchies, consisting of production nodes. Under conditions where it must deal with large volumes of requests, the core may span more than one level of the hierarchy, thus requiring vertical as well as horizontal ties. Within the core, everyone must support everyone in order not to be overwhelmed by external pressure; hence, the hierarchical substrate effectively disappears. In this scenario, the model yields a distinct class of information managers. Somewhat like the executives on the Delta Shuttle, these individuals spend all their time processing the information requests of production-oriented workers. And because their principal duty is to direct messages correctly, managers need to be highly connected to other managers (hence all the meetings).

Although it is an extreme configuration, too extreme perhaps for a

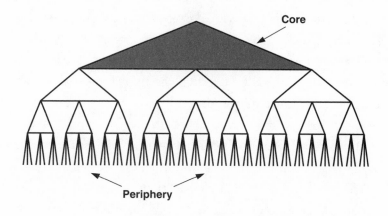

Figure 9.5. When message passing is purely global, congestion is concentrated at the top of the hierarchy, leading to a densely connected core comprising information managers, and a purely hierarchical periphery of specialized production workers.

human organization, this core-periphery architecture does bear some resemblance to the structure of mixed distribution/redistribution networks like the airline network and the postal system. Both systems consist of densely connected cores, within which passengers and letters, respectively, are redistributed and from which treelike distribution systems extend. In the airline network of the United States, for example, you can fly from any hub to almost any other directly; hence, the hubs form the core of the network. Each of the hubs then has its own local network of second- and third-tier airports to which it distributes passengers or receives them. The U.S. Postal Service also is partly a distribution system, collecting mail from many small outposts (post office boxes and minor post offices, for example) and distributing it in turn to individual homes and business. And it is partly a redistribution system as well. The redistribution function, however, is largely separate from the distribution function, occurring principally between major post offices and exchange centers.

To a lesser extent, the same core-periphery structure can be found in the structure of the Internet, which consists of a relatively dense backbone, within which individual routers are connected to many others and from which many treelike structures branch out, descending through increasingly local service providers, down to the level of individual users (the leaf nodes at which the branching tree terminates). Although not as clear as for the airline network, the resemblance between the core-periphery model and the real Internet makes sense: The bulk of data exchanges occur between widely separated users, as opposed to users of the same local ISP; thus, the burden of information *redistribution* is concentrated in the backbone.

Modern business and public sector organizations, however, experience a kind of ambiguity that is more complicated than either the purely-local or purely-global extreme. Furthermore, the nodes in these networks are people, not routers or offices, so simple distinctions like distribution versus redistribution are hard to apply. It is therefore a reassuring outcome of the congestion-reduction algorithm that the second of these observations follows naturally from the first. True ambiguity, it seems, necessitates problem solving, and thus communication, on all scales of the organization at once. Typically, the bulk of problem-solving activity that individuals undertake in even a fast-moving and complex environment occurs on the local scale—that is, within their immediate team of coworkers. Less routine problems, however, still crop up on a regular basis, and as the Honda example from before demonstrates, these require searching farther afield for the relevant information and resources. Often it may not be necessary to search any farther than a different team in the same department. But as we saw with the Toyota group, searches must occasionally extend even farther, beyond the same department, beyond the same division, or even outside the firm, where the required frequency of searches decreases with their range but never disappears altogether.

Essentially what our results indicate is that when organizations need to handle information processing on many distance scales at once, the network architecture required to handle the processing load must be connected on multiple scales as well. Although the likelihood that two individuals will have information relevant to each other's productivity decreases with their separation in the hierarchy, the number of such possible dependencies increases. Just as in social networks, many more people in a large organization are distant from you than close. The result, similar to Jon Kleinberg's observation in chapter 5, is that large volumes of information are flowing at all different levels of the hierarchy. Hence, bypass channels are needed not just at the local team scale (as in Figure 9.4) and not just at the global organizational scale (as in Figure 9.5), but at all scales at once. But because the hierarchy naturally concentrates information processing in its upper echelons, the distribution of messages being passed and the distribution of the resulting bypass links are not the same.

The intuitive picture looks something like Figure 9.6. Instead of a single, highly connected core at the top, now we have links extending throughout the hierarchy. Unlike the purely local teams of Figure 9.4, however, organizations designed to operate in truly ambiguous environments must contain "teams" at different scales. Near the bottom of the hierarchy, individuals mostly initiate messages rather than process them; hence, relatively few bypass links are required. Messages transmitted between distant nodes, meanwhile, must be processed further up in the hierarchy, requiring managers to interact not only with their immediate peers but also across vertical levels. The result is the emergence of *meta teams*, distributed groups that are not as densely connected as the teams in Figure 9.4 or the core in Figure 9.5, but that can distribute their information-processing burden across multiple scales instead of concentrating it in just one.

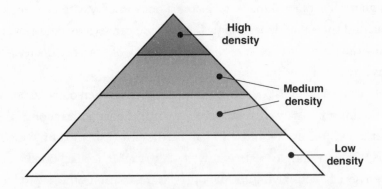

Figure 9.6. When message passing occurs at all scales, a multiscale network is required. The variegated shading reflects a decreasing density of ties with an increasing depth in the hierarchy.

A direct consequence of this *multiscale connectivity*, therefore, is that the distinction between a knowledge manager and a production worker becomes blurred. Although the tendency remains for processing activity to increase (at the expense of pure production) with an individual's height in the hierarchy, when information is being processed at all scales of the organization, everyone is to some extent managing information. The explanation for this breakdown in role distinctions is that in a truly ambiguous environment, where no one knows *exactly* what it is they are supposed to be doing, or how they are supposed to be doing it, problem-solving activity becomes inextricable from the task of production itself. Hence, everyone effectively must do some of both.

RECOVERING FROM DISASTER

LIKE AMBIGUITY, AN ORGANIZATIONAL FAILURE CAN COME IN many shapes and sizes—people get sick, factories burn down, computer systems crash, and large numbers of employees may have to be laid off. Sometimes the disaster comes from outside, and sometimes it is generated internally. Sometimes, as in the Aisin disaster, it is both: the fire was a natural event, but its significance was magnified by Aisin's exclusive production of brake valves and Toyota's just-in-time inventory system. Regardless of their origins, however, what all disasters have in common is that they disable a part of what was previously a whole, functioning system. Typically, in the long term, the disabled part must be repaired, replaced, or otherwise substituted for, say, by dispersing its functions permanently to other units. But in the fast-paced world of business, and in many physical networks like power grids and the Internet, surviving in the long run is not enough—first, the system must survive the short run.

As we saw with the Toyota-Aisin crisis, in the aftermath of any failure, the required rate of problem solving and information sharing dramatically increases. And when critical resources have been lost, the most important asset an organization can possess is easy access to the resources it has left. Hence, in network terms, the key to surviving a disaster in the short term is for the network to retain its connectivity while not incurring any more failures. This way of framing the problem puts us back in familiar terrain—sort of. Thinking about systemic robustness in terms of network connectivity is essentially the approach that Barabási and Albert introduced, and that Duncan Callaway later refined, to study the robustness of networks like the Internet. So to that extent, it's familiar. But all those results were based on the assumption that the networks at hand were random, and of course we are no longer talking about random networks.

Hierarchies, as you might imagine, perform terribly under conditions of failure. For the same reason that they are vulnerable to congestion-related failure (because they are too centralized), if any of a hierarchy's top nodes do fail, they will isolate large chunks of the network from each other. It is here that connectivity at all scales really comes into its own, for in multiscale networks there are no longer any "critical" nodes whose loss would disable the network by disconnecting it. And because they are designed to be decentralized not only at the level of teams but also at larger scales, they can survive bigger failures as well, such as if an entire team is taken out of commission. Essentially one can remove chunks of almost any size from a multiscale network and it will remain connected, and thereby able to access whatever resources were not directly destroyed.

Multiscale connectivity, therefore, serves not just one but two purposes that are essential to the performance of a firm in uncertain environments. By distributing the information congestion associated with problem solving across many scales of the organization, it minimizes the likelihood of failures. And *simultaneously* it minimizes the effect of failures if and when they do occur. Multiscale networks, therefore, satisfy the condition suggested at the end of chapter 8: that true robustness is a matter not only of avoiding failures but also of surviving them with a minimum of further loss. Because they exhibit this two-for-the-price-of-one robustness property,, we call multiscale networks *ultrarobust*.

Ultrarobustness might seem too good to be true, but actually it makes perfect sense, the key to conceptual development being that routine and exceptional problem-solving capabilities are intimately related. Routine ambiguity—uncertainty about what the world is going to throw at you tomorrow—drives routine problem solving. Routine problem solving, in turn, by generating information congestion at critical nodes, stimulates the creation of bypass links to relieve local congestion. And when information is being redistributed at all scales of the organization, bypasses get constructed at all scales as well. Once they exist, these multiscale links then have the additional property of keeping the network connected even in the event of large failures. The resulting resilience to external shocks, therefore, is an *unintentional* consequence of local coping mechanisms that individuals employ on a routine basis.

Another way to think about catastrophe recovery, therefore, is as a problem-solving activity, more dramatic than the routine kind of problem solving that firms have to face every day but fundamentally no different. Hence ultrarobustness is really nothing more then coping with ambiguity in all its manifestations. Sometimes the ambiguity is routine, and sometimes—as with the Toyota-Aisin crisis—it is extreme. But in all these cases, individuals are faced with an unfamiliar problem that they have to solve quickly. And the generic coping mechanisms that work for one problem often work for another. Innovation, error correction, and catastrophe recovery are thus all versions of what is essentially the same response to ambiguity.

Viewed in this light, the Toyota group's serendipitous recovery, even if not planned, was not serendipitous at all. Suddenly individuals found themselves confronted with a situation they had never imagined and probably would not have expected themselves to solve. Yet, given no choice but to deal with a dramatically altered reality, they solved it anyway, thus displaying collectively an organizational capability they never knew they had. But they *did* have it—it did not simply appear on demand. The workers of the Toyota group performed outstandingly in the face of a major crisis, but they were not supermen. Rather their crisis was simply a beefed up version of their routine response to everyday problems.

In the final analysis, ambiguity still exhibits an ambiguous character, but now it is one we can define and understand. On the one hand, chronic environmental ambiguity is the source of many of a firm's problems, continuously upsetting known routines and rendering current solutions obsolete. On the other hand, by making problem solving itself a routine activity, and by forcing organizations to build architectures that are capable of processing large volumes of information without failing, routine ambiguity is also a firm's best friend. By coping with everyday ambiguity, the firm develops the capability to rescue itself in the event of unanticipated disaster. Routine problem solving both balances the information-processing burden across the individuals of an organization *and* sets up the conditions under which exceptional problems can be resolved.

The precise *mechanism* by which a firm's response to routine ambiguity generates ultrarobustness is, as yet, an unsolved puzzle, but it seems to bear a deep resemblance to the property of network searchability that we encountered in chapter 5. To the extent that we understand it, the mechanism is something like the following. In order to resolve the problem of control, firms attempt to design themselves along hierarchical principles. But in ambiguous environments, information congestion related to problem-solving activities causes individuals—especially those higher in the hierarchy—to become overburdened. The local response of these individuals is to direct their subordinates to resolve problems on their own by conducting directed searches.

Lacking a central directory of organizational knowledge and resources, the subordinates rely on their informal contacts within their firm (or possibly in other firms) to locate relevant information. And as we know from chapter 5, this strategy of social search is effective; hence, searchers are able to succeed even though they may be unaware of precisely how or why their methods worked. Both the function and the structure of the hierarchy have now changed. Instead of rewarding just production performance, the chain of command now rewards search performance as well, so individuals have not only the capability to locate relevant information (a latent capacity) but also the incentive. A direct consequence is that the internal architecture of the firm is driven away from that of a pure hierarchy by virtue of the new links that are formed and consolidated over many repeated searches.

The equilibrium state of this process is a multiscale network for the simple reason that only when the network is connected across multiple scales is individual congestion—hence the pressure to create new connections—relieved. And as we saw in chapter 5, the presence of ties at multiple scales also renders the network highly searchable, so the multiscale state becomes self-reinforcing. When a catastrophe occurs, the network's properties of searchability and congestion relief appear serendipitous but are a natural consequence of local responses to routine problems in a chronically ambiguous environment.

The relationship between searchability and robustness, therefore, is a subtle one, manifesting itself as an intermediate case between our sociologically-motivated explanation of decentralized search in networks and Jon Kleinberg's explanation. Whereas Kleinberg's solution is completely *designed* in the sense that an engineer might design a circuit board from scratch, real organizations evolve largely as a consequence of locally informed decisions and individual searches. But unlike pure social networks, organizations do not evolve entirely spontaneously either. The hierarchy, in fact, is the signature of a designed solution to the internal organization of a firm. It happens to be a poor solution in the presence of ambiguity, but it is one that, as we have seen, can be modified to work well under a wide variety of circumstances.

Modern business firms, therefore, exploit the capacity for decentralized search that is latent in informal social networks by imposing on them the incentive structure inherent to hierarchies. Although we still don't fully understand the problem, it appears that a good strategy for building organizations that are capable of solving complex problems is to train individuals to react to ambiguity by searching through their social networks, rather than forcing them to build and contribute to centrally designed problem-solving tools and databases. The big payoff of this approach is that by understanding how individuals search socially, we can hope to design more effective *procedures* by which robust organizations can be constructed without having to specify the precise details of the organizational architecture itself.

CHAPTER TEN

The End of the Beginning

THE ISLAND OF MANHATTAN. TWENTY-TWO MILES LONG AND LESS than five miles wide, it is, on the grand scale of the world, a speck, a jewel in the mouth of the Hudson River as it pours into the North Atlantic. Up close, it is more like a vast roaring playground. Home to one and a half million people and host to millions more every day, it is, and has been for more than a century, Gotham, the quintessential metropolis, the city that never sleeps.

But from a scientific point of view, Manhattan is something of an enigma. Even on a daily basis, several million people, along with the private and commercial activity they generate, consume an awful lot of stuff—food, water, electricity, gas, and a huge range of materials from plastic wrapping to steel girders to Italian fashions. They also discharge an enormous quantity of waste in the form of garbage, recyclables, sewage, and waste water; collectively they emit so much raw heat energy that they create their own microclimate. Yet almost nothing that the city requires to sustain itself is actually produced or stored within its own precincts, nor can it satisfy any of its own disposal needs. Manhattan's drinking water is piped directly from the Catskill Mountains, two hours' drive upstate. Its power is generated as far away as the Midwest. And its food is trucked in from all over the nation and shipped in from around the world. Meanwhile, for decades its garbage has been transported by giant barges to the Fresh Kills landfill on nearby Staten Island, one of only two man-made constructions visible from space (the Great Wall of China being the other).

Another way to understand Manhattan, therefore, is as a nexus of flows, the swirling convergence of people, resources, money, and power. And if those flows stop, even temporarily, the city starts to die, starved for nourishment or choking on its own excrement. Grocery stores have the capacity to stock only days' worth of inventory—restaurants less. If the garbage is not picked up even once, it starts to pile up on the streets. And after the disastrous blackout of 1977, one cannot even imagine what would happen these days should power be cut off for more than a few hours. New Yorkers are renowned for their brash confidence, projecting an air of capability even in the most trying circumstances. But really they are captives of the very systems that make life in the city so convenient. From subway trains to bicycle deliverymen, from the water in the taps to the electricity that drives the elevators, every day they rely on the robust performance of an enormous complex infrastructure without which the trivial details of their lives—eating, drinking, commuting—would become unbearably onerous.

What would happen if this infrastructure, or even a part of it, were to stop functioning? Can it stop functioning? And who is in a position to ensure that it doesn't? *Who, in other words, is in charge?* Like many questions to do with complex systems, this one lacks a definitive answer, but the short version is *no one.* In reality, there is no such thing as a single infrastructure to be in charge of. Rather what exists is a Byzantine mishmash of overlapping networks, organizations, systems, and governance structures, mixing private and public, economics, politics, and society. The business of transporting people into, out of, and around Manhattan alone is divided among at least four distinct rail services, the subway system, dozens of bus companies, and several thousand taxis. Meanwhile, the multiple bridges and tunnels managed by the Port Authority, along with thousands of miles of roads and highways, enable literally millions of private vehicles to enter and leave the island every day. Food and mail are even more decentralized, involving hundreds of delivery services that collectively pump thousands of trucks, vans, and even bicycles around the streets of Manhattan, twenty-four hours a day, seven days a week.

No single entity coordinates this incredibly complicated, bewildering, impossible system, and no one understands it. Yet every day, when you stop by your local deli at 2 A.M. to pick up your favorite flavor of Ben & Jerry's ice cream, it's there, and someone is invariably stacking the brightly lit shelves with newly arrived produce of one sort or another. The system is a fact of life that the residents of Manhattan take for granted, but really it is a miracle that it works at all. And if this thought does not occasionally disturb their peace of mind, it really should. If there is one thing that we have learned from the previous chapters, it is that complex connected systems can display not only tremendous robustness in the face of adversity but also shocking fragility. And when the system is as complicated as a large, densely populated, heavily built-up city, as vital to the lives of millions of people, and as central to the economy of a global superpower, contemplating its potential break points is more than idle speculation. So how robust is Manhattan?

SEPTEMBER 11

On TUESDAY, SEPTEMBER 11, 2001, WE BEGAN TO FIND OUT. THE events of that terrible day, along with their social, economic, and political implications, have already been exhaustively analyzed. But there is reason to revisit the tragedy in the context of this story, as it illustrates many of the paradoxes we have encountered: how it is that connected systems can be at once robust and fragile; how apparently distant events can be closer than we think; how, at the same time, we can be insulated from what is happening even nearby; and how the routine can prepare us for the exceptional. The attacks of September 11 exposed, in a way that only true disasters can, the hidden connections in the complex architecture of modern life. And from that perspective, we still have some lessons to learn.

Purely from an infrastructural perspective, the attacks could actually have been worse. Unlike a nuclear blast or even the aerial release of a biological agent, the attack site was relatively localized, even somewhat isolated from the rest of the city. Far fewer transportation links, for instance, run through what was the World Trade Center than, say, Times Square or Grand Central Terminal. Nevertheless, the collapse of the towers dealt a massive physical blow, burying streets, crushing subway tunnels, and destroying one of the city's main telecommunication centers—the Verizon building at 140 West Street. Much of this damage will take years to repair, with even the direct cost estimated in the billions.

On that Tuesday, however, an equally significant consequence of the physical damage was that it precipitated a severe *organizational* crisis. The mayor's emergency command bunker was destroyed when 7 World Trade Center collapsed soon after the twin towers, and by 10 A.M. the nearby police command center had lost every single phone line, along with its cellular, e-mail, and pager services. Faced with a completely unexpected and unprecedented catastrophe, with almost no reliable information available and with the threat of subsequent attacks looming large, the city had to coordinate two enormous operations—one rescue and the other security—simultaneously. And less than an hour after the emergency began, the very infrastructure that had been designed to manage emergencies had been thrown into disarray.

But somehow the city did it. In what was, under the circumstances, an incredibly orderly response, the mayor's office, the police and fire departments, the Port Authority, the various state and federal emergency agencies, dozens of hospitals, hundreds of businesses, and thousands of construction workers and volunteers turned Lower Manhattan from a war zone into a recovery site in less than twenty-four hours. In the rest of the city, meanwhile, everything continued to operate in a way that was so normal, it was eerie. The power was still on, the trains still ran, and up at Columbia, you could still have a nice lunch at one of the restaurants on Broadway. For all the lockdown security on the island that day, nearly everybody outside the immediately devastated area got home that night, and deliveries of supplies and collection of garbage returned to almost normal the next day. The police still patrolled the city, and the fire department, despite losing more than twice as many men in one hour as are normally lost in the entire country in an entire year, still responded to every alarm. That night, friends watched the president's televised address in bars that were as crowded as ever, and the next day most of the city actually went back to work. The daily routines of life persisted to such an extent, in fact, that many New Yorkers felt guilty for not being affected *enough*.

By Friday, the barriers cordoning off the southern tip of the island had retreated south from 14th Street to Canal, and by the following Monday, September 17, most of the downtown area was ready for business again. Even the stock exchange reopened, despite the financial industry having borne the brunt of the enormous losses in both personnel and materiel. From private homes, shared offices, and borrowed floor space all over Manhattan, Brooklyn, New Jersey, and Connecticut, trading firms desperately scrambled to reconfigure themselves, salvaging data from backup servers, jury-rigging temporary communication systems, and struggling not only to cope with but also to compensate for the tragedy of lost colleagues.

Morgan Stanley alone had thirty-five hundred people working in the south tower. Incredibly none of them were lost, but that didn't reduce the problem of relocating thousands of people within a matter of days, when for much of that time it wasn't clear how many of them were even alive! Many other companies, large and small, faced a similarly daunting task. Merrill Lynch, for example, housed across the street in the World Financial Center, didn't lose its offices but still had to relocate several thousand workers for more than six months, until they could reenter the building. All told, more than one hundred thousand people had to find somewhere else to go to work that Monday. A troop deployment of this magnitude would be almost unfathomable with less than a week's warning, even for the army, which is designed specifically for such an exercise. Yet somehow, at 9:30 Monday morning, a mere six days after the world had seemed to end, the opening bell of the New York Stock Exchange rang again.

As with the Toyota-Aisin crisis, all the firms and government agencies involved in the recovery effort certainly had strong incentives—economic, social, and political—to do what they did. But as pointed out in chapter 9 even the strongest incentives are not sufficient to generate an effective response in the short term—the capability had to be there also. And as with the Toyota group, the capability to recover from the catastrophe could have not been consciously designed. In fact, to the extent that it *had* been designed—the mayor's emergency command center, for example—it mostly failed to work, or at least not as intended. Nor was there sufficient time, in the midst of the crisis itself, for all the parties involved to learn everything they needed to know. So whatever it was about the system that enabled it to recover so rapidly had to have been there beforehand and had to have evolved principally for other purposes.

A few months after September 11, I heard a remarkable story told by a woman from Cantor Fitzgerald, the debt trading firm that lost more than seven hundred of its roughly one thousand employees in the collapse of the north tower. Despite (or perhaps because of) the unfathomable trauma they had just suffered, the remaining employees decided by the next day that they would try to keep the firm alive, a decision made all the more incredible by the daunting practical hurdles they needed to overcome. First, unlike the equity markets, the fixed-income markets were not based at the Stock Exchange and had not closed. So if it was to survive, Cantor Fitzgerald needed to be up and running within the next forty-eight hours. Second, while their carefully constructed contingency plan had called for remote backups of all their computer and data systems, there was one eventuality they had not anticipated: *every single person who knew the passwords had been lost.* And the reality is that if no one knows the passwords, the data are as good as gone, at least on the timescale of two days.

So what they did is this: They sat around in a group and recalled everything they knew about their colleagues, everything they had done, everywhere they had been, and everything that had ever happened between them. *And they managed to guess the passwords.* This story seems hard to believe, but it's true. And it illustrates, in a particularly dramatic way, the point from the last chapter: recovery from disaster is not something that can be planned for in an event-specific manner, nor can it be centrally coordinated at the time of the disaster itself. Just as with the mayor's office, and the Aisin company before it, in a true disaster the center is the first part of the system to be overwhelmed. The system's survival, therefore, like Cantor Fitzgerald's, depends on a distributed network of preexisting ties and ordinary routines that binds an organization together across all its scales.

What was really so remarkable then about the robustness of downtown New York was that the survival and recovery mechanisms used by people, firms, and agencies alike were not remarkable at all. After all the fancy electronics of the mayor's emergency command had been disabled, the burden of communication fell on police radios and pieces of paper ferried back and forth by speeding patrol cars. In the absence of clear directions, paramedics, construction workers, off-duty firefighters, and volunteers simply showed up and were quickly assimilated into a routine that largely crystallized along the lines on the ground, not acording to any preconceived design. And the scattered survivors of Cantor Fitzgerald found each other by walking to each other's homes. In the immediate aftermath, it is worth remembering, nobody knew what was going on—neither the troops nor the generals—and nobody knew how they were supposed to respond. So they did the only thing they could do: they followed their routines and adapted them as best they could to allow for the dramatically altered circumstances. In some cases, this strategy was disastrous—the firefighters who rushed up the stairwells to their doom were also following their routines—but mostly it worked surprisingly well. "Ordinary heroes" was a descriptor applied over and over again in the months after September 11. But from an organizational perspective, what we should learn from the recovery effort is that the exceptional is really all about the routine.

Six months later, however, the fragile side of the same system had been revealed, with virtually every industry, from insurance and health care, to transportation, entertainment, tourism, retail, construction, and finance, being affected in some adverse way by the attacks. A number of restaurants in Lower Manhattan went out of business almost immediately, after being forced to shutter their doors for days or even weeks, and several Broadway shows closed due to a sag in audience levels. Within a month, thousands of workers had been laid off in the financial industry, and most of the rest forfeited their annual bonus, thus effectively cutting their pay by as much as 75 percent. Although the financial sector accounts for only 2 percent of the jobs in New York, it generates nearly 20 percent of the city's income, so cuts of that magnitude have the potential to reverberate all the way up the island, affecting not just retail and rents but even public revenues that are used to clean the streets, make the subways safe, and keep the parks beautiful.

Even worse, the main reason why many financial firms were located in Lower Manhattan was essentially because so many other firms were *already* there. In the last decade or so, however, as financial transactions have become increasingly electronic, and physical proximity has become less and less relevant, some firms have begun to drift away. Now that the trade center is gone, and many companies are faced with relocation decisions *simultaneously*, that drift may become a stampede. If it does, much of the associated revenue on which New York has come to depend may go elsewhere, returning the city to the fiscal doldrums of the 1970s. No one yet knows the likelihood of this dismal scenario, and plenty of more optimistic alternatives have been proposed. The point is not to make particular predictions, but to highlight that the city is connected in ways that are hard to anticipate and even harder to direct.

The connections, of course, do not stop at the Hudson River. The effects of the attacks have been felt on a national scale too. Midway Airlines (based in North Carolina) declared bankruptcy the day after the attacks, and by the end of the week, almost every national carrier had indicated severe financial distress. Over one hundred thousand airline workers were eventually laid off. The country's economy, already on the brink of recession, looked ready to implode if investors fled U.S. investments and consumers failed to pick up the slack. Although with the economy now staging a weak recovery and the more pessimistic projections seeming unlikely, the collateral damage has still been significant. After a disappointing Christmas season, one of the nation's largest retailers, Kmart, filed for bankruptcy protection, leaving a mountain of unresolved debt, which in turn might trigger additional bankruptcies of unpaid creditors.

What are we to conclude? Were the attacks more damaging than they initially seemed, or less? Did the system respond robustly, or were its hidden vulnerabilities exposed? In a thought-provoking but ultimately frustrating piece in the *New York Times* a few weeks later, the economist Paul Krugman aired his speculations about the impact of the attacks on an already weakening U.S. economy. As usual, Krugman's arguments were lucid, impeccably reasoned, and compelling. But effectively all he said was that there were a number of good reasons why the U.S. economy would rebound and be fine over the foreseeable future. And there were an equal number of equally plausible reasons why the U.S. economy would spiral into a long-term, damaging recession. He didn't want to say he had no idea what was going to happen (and his artful hedging enabled him to claim a correct prediction regardless of the outcome), but it was pretty clear he didn't. And Krugman is one of the world's best economists, especially when it comes to the business of explaining real economic phenomena. So if Krugman and his Princeton buddies have no idea how complex economic systems respond to large shocks, it's a fair bet that nobody does.

What can the science of networks tell us that Krugman can't? The honest answer, unfortunately, is not too much—yet. It is important to recognize that despite fifty years of percolating in the background, the science of networks is only just getting off the ground. If this were structural engineering, we would still be working out the rules of mechanics, the basic equations governing the bending, stretching, and breaking of solids. The kind of applied knowledge to which professional engineers have access—the tables, handbooks, computer design packages, and heavily tested rules of thumb—is at best on the distant horizon. But what the science of networks can do is provide a new way of thinking about familiar problems, a way that has already yielded some surprising insights.

LESSONS FOR A CONNECTED AGE

First, the science of networks has taught us that distance is deceiving. That two individuals on the opposite sides of the world, and with little in common, can be connected through a short chain of network ties—through only *six degrees*—is a claim about the social world that has fascinated generation after generation. The explanation, as we saw in chapter 3, derives from the existence of social connections that span long distances, and from the fact that only a few such ties can have a big impact on the connectedness of the whole world. As we then saw in chapter 5, the origin of these long-distance links resides in the multidimensional nature of social identity—we tend to associate with people like ourselves, but we have multiple independent ways of being alike. And because we know not only who our friends are, but also what kind of people they are, even very large networks can be navigated in only a few links.

But even if it is true that everyone can be connected to everyone else in only six degrees of separation, so what? How far is six degrees anyway? From the point of view of getting a job, locating information, or getting yourself invited to a party, anyone more distant than a friend of a friend is, for all intents and purposes, a stranger. So as far as extracting resources is concerned, or exerting influence, anything more than two degrees might as well be a thousand. We may be connected, but that doesn't make us any less foreign to each other, nor does it necessarily incline us to reach out beyond the little clusters that define our individual lives. In the end, we all have our own burdens to bear, and to concern ourselves with the multitudes of distant others would only drive us crazy.

But sometimes those distant multitudes have a way of showing up on our doorstep, uninvited. In 1997, the decoupling of the Thai bhat from the U.S. dollar triggered a real estate crisis in Thailand that led to the collapse of its banking system. Within months, financial distress had spread to the other "Asian tigers"—Indonesia, Malaysia, and South Korea—deflating their previously red-hot economies and prompting a global depression in the prices of commodities, especially oil. Russia, meanwhile, in the throes of a traumatic transition to a capitalist economy, was heavily dependent on its oil exports, and suddenly its precious black gold wasn't worth as much. A Russian budgetary crisis ensued, and the government was forced to default on its sovereign debt, not something that even former superpowers are supposed to do. The shock to the world debt markets caused investors to flee bonds of any kind other than those of the U.S. government.

Just prior to this, and little known to the rest of world, a hedge fund in Greenwich, Connecticut, called Long Term Capital Management (LTCM), had placed huge bets on what it perceived were mispricings in a wide range of bonds. But now, to the fund's horror, the prices that were supposed to converge began to diverge instead, evaporating positions worth billions of dollars in a few months. Concerned that if LTCM were forced to liquidate its assets, the very markets in which it operated might fail, the chairman of the New York Federal Reserve coordinated a bailout by a consortium of the largest investment banks in the country, thus averting a potential catastrophe. And there the tidal wave that had crashed across Asia the year before stopped, lapping gently on the shores of Long Island Sound.

The United States largely escaped the 1997 Asian crisis, but no one knew what to expect at the time. Nor do they know what to expect this time around, with the religious and political turmoil of the Middle East manifesting itself as terror in the skies over New York and Washington, D.C. In a world spanned by only six degrees, what goes around comes around faster than you think. So just because something seems far away, and just because it happens in a language you don't understand, doesn't make it irrelevant. When it comes to epidemics of disease, financial crises, political revolutions, social movements, and dangerous ideas, we are all connected by short chains of influence. It doesn't matter if you know about them, and it doesn't matter if you care, they will have their effect anyway. To misunderstand this is to misunderstand the first great lesson of the connected age: we may all have our own burdens, but like it or not, we must bear each other's burdens as well.

The second major insight we can gain from the science of networks is that in connected systems, cause and effect are related in a complicated and often quite misleading way. Sometimes small shocks can have major implications (see chapter 8), and other times even major shocks can be absorbed with remarkably little disruption (see chapter 9). This point is extremely important because most of the time we are only in a position to judge the significance of things in retrospect, and in retrospect it is easy to be wise. After the first *Harry Potter* book became an international phenomenon, everyone rushed to heap praise on its quality as a children's novel, and every subsequent installment instantly became a best-seller. Probably the success of the series is thoroughly deserved, but what we forget is that several publishers rejected J. K. Rowling's original manuscript before it was picked up by Bloomsbury (which at the time, was a small, independent press). If the quality of her work is so obvious, why wasn't it apparent to numerous experts in the children's book publishing industry? And what does that suggest about all the other rejected manuscripts languishing in various editors' drawers around the world? In 1957, Jack Kerouac's *On the Road* became an American classic almost overnight. But what few of its inspired readers realize is that it almost failed to see the light of day: Kerouac completed the original manuscript six years before Viking agreed to publish it. What if he had given up? After all, many authors do. How many classics has the world lost as a result?

On the converse, what if the Toyota group had not found a way to cope with the Aisin disaster? This scenario is perfectly conceivable. Major companies do go out of business—Enron and Kmart being but two recent examples—and the potential disruption to Toyota's business easily could have been large enough to force it into bankruptcy. What effect would that have had? Had the world suddenly been deprived of its beloved Toyota vehicles, the Aisin disaster would have made headlines for months. And by sealing the fate of not just Toyota but possibly many of its approximately two hundred suppliers, it may also have caused a severe meltdown in an already depressed Japanese economy. That would have made it one of the biggest stories of the decade. As it is, however, no one outside of a few industrial organization specialists has even heard of the Aisin crisis. Because it had quite limited *consequences* for the global economy, it has become an historical footnote. But the results could so easily have been different. Much the same comment applies (although for completely different reasons) to the Ebola outbreak among the monkeys in Reston, Virginia, described in chapter 6. What if the virus *had* been Ebola Zaire? The United States might well have experienced a major public health catastrophe on the doorstep of its capital. Yet the only reason we even know about it is because Richard Preston wrote such an engaging book (and found a good publisher!).

History, therefore, is an unreliable guide to an unpredictable future. We rely on it anyway, because it seems like we have no other option. But maybe we do have another option—not for predicting specific outcomes perhaps, but for understanding the mechanisms by which they manifest themselves. And sometimes understanding can be enough. Darwin's theory of natural selection, for instance, doesn't actually predict anything. Nevertheless, it gives us enormous power to make sense of the world we observe, and therefore (if we choose) to make intelligent decisions about our place in it. In the same way, we can hope that the new science of networks can help us understand both the structure of connected systems and the way that different sorts of influences propagate through them.

Already we can understand that connected, distributed systems, from power grids to business firms to even entire economies, are *both* more vulnerable and more robust than populations of isolated entities. If two individuals are connected by a short chain of influences, then what happens at one *may* affect the other even if they are completely unaware of each other. If that influence is damaging, then each is more vulnerable than they would be if they were alone. On the other hand, if they can find each other through that same chain, or if they are both embedded in some mutually reinforcing web of relations with other individuals, then each may be capable of weathering a greater storm than they could by themselves. Networks share resources and distribute loads, and they also spread disease and transmit failure—they are both good and bad. By specifying precisely *how* connected systems are connected, and by drawing explicit relationships between the structure of real networks and the behavior (like epidemics, fads, and organizational robustness) of the systems they connect, the science of networks can help us understand our world.

And finally, the science of networks is showing us that it really is a new science, not one that belongs as a subset of any traditional scientific endeavor but one that crosses intellectual boundaries and draws on many disciplines at once. As we have seen, the mathematics of the physicists is paving new roads through previously unexplored terrain. Random growth, percolation theory, phase transitions, and universality are the bread and butter of physicists, and they have found a wonderfully open set of problems in networks. But without the maps of sociology, economics, and even biology to guide them, the physicists are quite capable of building roads to nowhere. Social networks are not lattices, and not everything can be scale-free. One kind of percolation works for some problems but not for others. Some networks are built on hierarchies and some aren't. In some respects the behavior of the system is independent of the particulars, but some details still matter. For any complex system, there are many simple models we can invent to understand its behavior. The trick is to pick the right one. And that requires us to think carefully—*to know something*—about the essence of the real thing.

To belabor the point a bit more, claiming that everything is a small-world network or a scale-free network not only oversimplifies the truth but does so in a way that can mislead one to think that the same set of characteristics is relevant to every problem. If we want to understand the connected age in any more than a superficial manner, we need to recognize that different classes of networked systems require us to explore different sorts of network properties. In some cases, it may be sufficient to know simply that a network contains a short path connecting any pair of individuals, or that some individuals are many times better connected than others. But in other cases, what may matter is whether or not the short paths can be found by the individuals themselves. Perhaps it may be important that in addition to being connected by short paths, individuals are also embedded in locally reinforcing clusters, or that they are not so embedded. Sometimes the existence of individual identity may be critical to understanding a network's properties, and other times it may not be. Being highly connected may be of great use in some circumstances and of little consequence in others—it may even be counterproductive, leading to failures or exacerbating failures that occur naturally. Just like the taxonomy of life, a *useful* taxonomy of networks will enable us both to unify many different systems and to distinguish between them, depending on the particular questions we are asking.

Building the science of networks, therefore, is a task that will require a major coordinated push by all the disciplines and even the professions, bringing to the struggle the mathematical sophistication of the physicist, the insight of the sociologist, and the experience of the entrepreneur. It's a monumental task, and sometimes, I have to say, it seems hopeless. We struggle so long to learn so little, it's tempting to believe that the connected age is just too complicated to be understood in any kind of systematic, scientific way. Perhaps, despite all our best efforts, we will ultimately have to content ourselves with the role of observing the inscrutable and intractable game of life, relegated merely to getting up each morning and seeing what happens. But no one is giving up yet.

Perhaps the most inspiring aspect of science is that its very nature is to ask questions that do not yet have answers. In that sense, science is a fundamentally optimistic exercise. Not only do scientists persistently believe in the comprehensibility of the world, but also they are unde-terred by the ultimate limitations on what they can do. Beyond any problem, however difficult, lies one that is harder still, and no level of understanding is ever complete. Every disease cured brings another to the fore. Every invention has unintended consequences. And every successful theory merely heightens our standards for explanation. On bad days, every scientist feels a bit like Sisyphus, endlessly rolling his rock up the hill, only to wind up at the bottom the next day. But Sisyphus *kept going* and so it is with science—even when it seems hope-less, we struggle on, because, as with most human ambitions, it is in the struggle that we find the measure of ourselves.

Besides, just because the mysteries of the connected age often seem incomprehensible doesn't mean they are. Before Copernicus, Galileo, Kepler, and Newton, the motions of the heavenly bodies were consid-ered transparent to the mind of God alone. Before Orville and Wilbur Wright flew their first plane at Kitty Hawk, man wasn't meant to fly. And before a climber named Warren Harding struggled three thousand desperate feet up the granite face of El Capitán, it was thought that no man could climb it. In every field of human endeavor, there is always the impossible. And in every field there are always those who would attempt it. Most of the time they fail, and the impossible remains just that. But once in a while, they succeed, and it is at these jump points that we collectively pass to the next level of the great game.

Science is not a domain renowned for its heroes, there being little that is glamorous about a scientist's everyday business—frankly, it does not make for good TV. But every day, scientists pit themselves against the impossible, struggling to understand the very parts of the world that are not—and never have been—understood. The science of networks is just one skirmish in this many-fronted conflict. But it is one that is rapidly gaining the attention of the broader scientific community. And more than fifty years after the likes of Rapoport and Erdős fired the first shots, it may be that the battle is starting to turn in our favor. Or as Winston Churchill said after the battle of El Alamein in 1942, "This is not the end. It is not even the beginning of the end. But it is, perhaps, the end of the beginning."

Further Reading

The following pages are designed as a guide for readers who want to expand their understanding of the science of networks, start doing their own research, or simply learn some more details about the topics mentioned in the book. As such, the readings are coded by difficulty, according to the simple rating scale given below. Also, the readings are broken down by chapter and section, thus enabling a quick association between specific topics and the relevant material. A complete, alphabetized bibliography follows. As long as the reading list is, it is far from comprehensive. In fact, most subjects have been covered rather sparsely, if at all, and I apologize to any authors that I have erroneously excluded. Such obvious omissions notwithstanding, it remains the case that by starting here, the uninitiated reader can quickly find much of the major literature associated with the new science of networks. And a lot more besides.

DIFFICULTY RATING
- **Beginner** (no more difficult than this book)
- **Intermediate** (some effort and mathematical background required)
- **Advanced** (requires undergraduate-level mathematics training)
- **Expert** (impenetrable without graduate-level mathematics training)

CHAPTER ONE: THE CONNECTED AGE

The proximate causes and immediate effects of the cascading failure that occurred in the Western Systems Coordinating Council transmission grid on August 10, 1996, are described in
• WSCC Operations Committee. *Western Systems Coordinating Council Disturbance Report, August 10, 1996* (October 18, 1996). Available on-line at http://www.wscc.com/outages.htm.

A review of major recent power disturbances in the United States that sheds some additional light on the August 10 cascade is
• Hauer, J. F., and Dagel, J. E. *White Paper on Review of Recent Reliability Issues and System Events.* Consortium for Electric Reliability Technology Solutions. U.S. Department of Energy (1999). Available on-line at http://www.eren.doe.gov/der/transmission/pdfs/reliabilityevents.pdf.

Some theoretical papers dealing with the problem of cascading failures in power transmission networks are
■ Kosterev, D. N., Taylor, C. W., and Mittelstadt, W. A. Model validation for the August 10, 1996 WSCC system outage. *IEEE Transactions on Power Systems,* 14(3), 967–979 (1999).
◆ Sachtjen, M. L., Carreras, B. A., and Lynch, V. E. Disturbances in a power transmission system. *Physical Review E,* 61(5), 4877–4882 (2000).
◆◆ Asavathiratham, C. The influence model: A tractable representation for the dynamics of networked Markov chains (Ph.D. dissertation, Department of Electrical Engineering and Computer Science, MIT, 2000).

Emergence

Although the author doesn't call it emergence, one of the earliest works to really grapple with the issue of emergent behavior in complex (social) systems is
• Schelling, T. C. *Micromotives and Macrobehavior* (Norton, New York, 1978).

The classic paper by Phillip Anderson that outlines the fundamental concept of emergence is
■ Anderson, P. W. More is different. *Science,* 177, 393–396 (1972).

Some very readable introductions to the subject of complex, adaptive systems in general, and emergence in particular, are
- Gell-Mann, M. *The Quark and the Jaguar: Adventures in the Simple and the Complex* (W. H. Freeman, New York, 1994).
- Holland, J. H. *Hidden Order: How Adaptation Builds Complexity* (Perseus, Cambridge, MA, 1996).
- Waldrop, M. M. *Complexity: The Emerging Science at the Edge of Order and Chaos* (Touchstone, New York, 1992).

A more technical read is
- Casti, J. L. *Reality Rules I & II: Picturing the World in Mathematics: The Fundamentals, the Frontier* (Wiley-Interscience, New York, 1997).

Networks

A good introduction to the mathematical theory of graphs (and one that explains Euler's theorem in detail) is
- West, D. B. *Introduction to Graph Theory* (Prentice-Hall, Upper Saddle River, NJ, 1996).

A more applied approach to the subject, focusing more on algorithms and applications than on theorems, is presented in the following general texts:
- Lynch, N. A. *Distributed Algorithms* (Morgan Kauffman, San Francisco, 1997).
- Ahuja, R. K., Magnanti, T. L., and Orlin, J. B. *Network Flows: Theory, Algorithms, and Applications* (Prentice-Hall, Englewood Cliffs, NJ, 1993).
- Nagurney, A. *Network Economics: A Variational Inequality Approach* (Kluwer Academic, Boston, 1993).

Synchrony

The best way to learn about the subject of coupled oscillators is from Steve Strogatz himself in his recent book:
- Strogatz, S. H. *Sync: The Emerging Science of Spontaneous Order* (Hyperion, Los Angeles, 2003).

Strogatz has also written two shorter accounts of his (and related) work on the Kuramoto model:
- Strogatz, S. H. Norbert Wiener's brain waves. In Levin, S. A. (ed.),

Frontiers in Mathematical Biology, Lecture Notes in Biomathematics, 100 (Springer, New York, 1994), pp. 122–138.

- Strogatz, S. H., and Stewart, I. Coupled oscillators and biological synchronization. *Scientific American,* 269(6), 102–109 (1993).

The Road Less Traveled

Winfree's original paper on the entrainment of coupled oscillators that kicked off much of the recent literature, and that was my initial reference point, is

- Winfree, A. T. Biological rhythms and the behavior of populations of coupled oscillators. *Journal of Theoretical Biology,* 16, 15–42 (1967).

For those who want to learn more from the master, a fascinating (albeit a somewhat challenging) read is

- Winfree, A. T. *The Geometry of Biological Time* (Springer, Berlin, 1990).

The Small-World Problem

The famous *Psychology Today* paper that everyone cites when they talk about Milgram's work is

- Milgram, S. The small world problem. *Psychology Today,* 2, 60–67 (1967).

A better reference, however, is a paper that Milgram published two years later with his graduate student, Jeffrey Travers. It contains considerably more detail than the *Psychology Today* article, and although it isn't quite as fun, it is actually much clearer.

- Travers, J., and Milgram, S. An experimental study of the small world problem. *Sociometry,* 32(4), 425–443 (1969).

The very first study of the small-world problem was by Manfred Kochen and Ithiel de Sola Pool, who circulated their results as a working paper almost ten years before Milgram conducted his experiment. In fact, it was this paper that stimulated Milgram's work in the first place. It was finally published, almost twenty years after its inception, as the first paper in the first volume of the journal *Social Networks.*

■ Pool, I. de Sola, and Kochen, M. Contacts and influence. *Social Networks*, 1(1), 1–51 (1978).

A number of subsequent papers, both theoretical and empirical, as well as the original paper by Kochen and Pool, are available in a collected volume, edited by Manfred Kochen.

■ Kochen, M. (ed.). *The Small World* (Ablex, Norwood, NJ, 1989).

The play by John Guare that was later made into a movie, and launched the phrase "six degrees of separation" into the pop-culture stratosphere, is

• Guare, J. *Six Degrees of Separation: A Play* (Vintage Books, New York, 1990).

CHAPTER TWO: THE ORIGINS OF A "NEW" SCIENCE

The Theory of Random Graphs

Random-graph theory is not for the faint of heart. As a result, there are not really any readings one might call "accessible." Nevertheless, here are some important ones. The original results of Erdős and Rényi dealing with the evolution and connectedness of random graphs are contained in the following series of classic papers (none of which are easy to find in your average library):

◆◆ Erdős, P., and Rényi, A. On random graphs. *Publicationes Mathematicae*, 6, 290–297 (1959).

◆◆ Erdős, P., and Rényi, A. On the evolution of random graphs. *Publications of the Mathematical Institute of the Hungarian Academy of Sciences*, 5, 17–61 (1960).

◆◆ Erdős, P., and Rényi, A. On the strength and connectedness of a random graph. *Acta Mathematica Scientia Hungary*, 12, 261–267 (1961).

The standard textbook for the field of random graphs, and one that summarizes the most significant developments since Erdős and Rényi, is

◆◆ Bollobas, B. *Random Graphs*, 2d ed. (Academic, New York, 2001).

A slightly more readable (although not as comprehensive) treatment of the subject is

◆ Alon, N., and Spencer, J. H. *The Probabilistic Method* (Wiley-Interscience, New York, 1992).

Social Networks

The standard textbook for social network analysis is
- Wasserman, S., and Faust, K. *Social Network Analysis: Methods and Applications* (Cambridge University Press, Cambridge, 1994).

Two shorter, less comprehensive, but more readable alternatives are
- Degenne, A., and Forse, M. *Introducing Social Networks* (Sage, London, 1999).
- Scott, A. *Social Network Analysis*, 2d ed. (Sage, London, 2000).

Finally, a very short collection of classic papers—the ones that introduced some of the central concepts in the field—includes
- Boorman, S. A., and White, H. C. Social structure from multiple networks. II. Role structures. *American Journal of Sociology*, 81(6), 1384–1446 (1976).
- Burt, R. S. *Structural Holes: The Social Structure of Competition* (Harvard University Press, Cambridge, MA, 1992).
- Davis, J. A. Structural balance, mechanical solidarity, and interpersonal relations. *American Journal of Sociology*, 68(4), 444–462 (1963).
- Freeman, L. C. A set of measures of centrality based on betweenness. *Sociometry*, 40, 35–41 (1977).
- Granovetter, M. S. The strength of weak ties. *American Journal of Sociology*, 78, 1360–1380 (1973).
- Harary, F. Graph theoretic measures in the management sciences. *Management Science*, 5, 387–403 (1959).
- ◆◆ Holland, P. W., and Leinhardt, S. An exponential family of probability distributions for directed graphs. *Journal of the American Statistical Association*, 76, 33–65 (1981).
- Lorrain, F., and White, H. C. Structural equivalence of individuals in social networks. *Journal of Mathematical Sociology*, 1, 49–80 (1971).
- ◆◆ Pattison, P. *Algebraic Models for Social Networks* (Cambridge University Press, Cambridge, 1993).
- White, H. C., Boorman, S. A., and Breiger, R. L. Social structure from multiple networks. I. Blockmodels of roles and positions. *American Journal of Sociology*, 81(4), 730–780 (1976).

The Dynamics Matters

Because the subject of networks and dynamics is such a new one, there is really no text on the subject. One starting place is the following edited volume—a collection of roughly forty papers with introductions written by the editors:
◆◆ Newman, M. E. J., Barabási, A. L., and Watts, D. J. *The Structure and Dynamics of Networks* (Princeton University Press, Princeton, NJ, 2003).

A great introduction to the field of nonlinear dynamics is
■ Strogatz, S. H. *Nonlinear Dynamics and Chaos with Applications to Physics, Biology, Chemistry, and Engineering* (Addison-Wesley, Reading, MA, 1994).

And a discussion of its relevance to networks is
■ Strogatz, S. H. Exploring complex networks. *Nature,* 410, 268–275 (2001).

A paper that points out the limitations of centrality as a surrogate for social influence is
■ Mizruchi, M. S., and Potts, B. B. Centrality and power revisited: Actor success in group decision making. *Social Networks,* 20, 353–387 (1998).

The story quoting Jon Kleinberg on the connection to Bill Gates is
● Wildavsky, B. Small world, isn't it? *U.S. News and World Report,* April 1, 2002, p. 68.

And the article that suggested OTPOR as a good example of decentralized action is
● Cohen, R. Who really brought down Milosevic? *New York Times Magazine,* November 26, 2000, p. 43.

Departing from Randomness

Over a period of more than a decade, Anatol Rapoport produced a series of papers that laid out the theory of random-biased nets. The central ideas, however, are contained in the following pair of papers:
◆ Solomonoff, R., and Rapoport, A. Connectivity of random nets. *Bulletin of Mathematical Biophysics,* 13, 107–117 (1951).

◆   Rapoport, A. A contribution to the theory of random and biased nets. *Bulletin of Mathematical Biophysics*, 19, 257–271 (1957).

A summary of the random-biased network approach is given in
◆   Rapoport, A. Mathematical models of social interaction. In Luce, R. D., Bush, R. R., and Galanter, E. (eds.), *Handbook of Mathematical Psychology*, vol. 2 (Wiley, New York, 1963), pp. 493–579.

Rapoport's own account of his life, and his life's work, is
●   Rapoport, A. *Certainties and Doubts: A Philosophy of Life* (Black Rose Press, Montreal, 2000).

Here Come the Physicists . . .

The classic text on the theory of critical phenomena is
◆   Stanley, H. E. *Introduction to Phase Transitions and Critical Phenomena* (Oxford University Press, Oxford, 1971).

A more contemporary version is
◆   Sornette, D. *Critical Phenomena in Natural Sciences* (Springer, Berlin, 2000).

A detailed discussion of spin systems and phase transitions is given in
◆   Palmer, R. Broken ergodicity. In Stein, D. L. (ed.), *Lectures in the Sciences of Complexity*, vol. I, Santa Fe Institute Studies in the Sciences of Complexity (Addison-Wesley, Reading, MA, 1989), pp. 275–300.
◆   Stein, D. L. Disordered systems: Mostly spin systems. In Stein, D. L. (ed.), *Lectures in the Sciences of Complexity*, vol. I, Santa Fe Institute Studies in the Sciences of Complexity (Addison-Wesley, Reading, MA, 1989), pp. 301–354.

If one actually wants to do work in this field, a useful text is
◆   Newman, M. E. J., and Barkema, G. T. *Monte Carlo Methods for Statistical Physics* (Clarendon Press, Oxford, 1999).

And finally, a very accessible text that uses simple computer models to explain many of the central concepts of nonlinear dynamics and critical phenomena is

- Flake, G. W. *The Computational Beauty of Nature: Computer Explorations of Fractals, Chaos, Complex Systems, and Adaptation* (MIT Press, Cambridge, MA, 1998).

CHAPTER THREE: SMALL WORLDS

The original account of my work with Steve Strogatz was my Ph.D. dissertation, later published as

- Watts, D. J. *Small Worlds: The Dynamics of Networks between Order and Randomness* (Princeton University Press, Princeton, NJ, 1999).

With a Little Help from My Friends

A discussion of *agency,* as the term is understood by sociologists, is

- Emirbayer, M., and Mische, A. What is agency? *American Journal of Sociology,* 103(4), 962–1023 (1998).

The computer algorithms used to do the relevant calculations are quite standard and can be learned from any good text on algorithms. Two good examples are

- Aho, A. V., Hopcroft, J. E., and Ullman, J. D. *Data Structures and Algorithms* (Addison-Wesley, Reading, MA, 1983).
- Ahuja, R. K., Magnanti, T. L., and Orlin, J. B. *Network Flows: Theory, Algorithms, and Applications* (Prentice-Hall, Englewood Cliffs, NJ, 1993).

From Cavemen to Solarians

The two books from Asimov's *Robot* series that inspired my discussion with Steve are

- Asimov, I. *The Caves of Steel* (Doubleday, Garden City, NY, 1954).
- Asimov, I. *The Naked Sun* (Doubleday, Garden City, NY, 1957).

Small Worlds

The derivation of the alpha model, and the consequent identification of small-world networks, are presented in

- Watts, D. J. Networks, dynamics and the small-world phenomenon. *American Journal of Sociology,* 105(2), 493–527 (1999).

A simpler, but similar, version of the model was later studied by
◆ Jin, E. M., Girvan, M., and Newman, M. E. J. The structure of growing networks. *Physical Review E, 64,* 046132 (2001).

As Simple as Possible

The beta model and the empirical results about small-world networks were first published in
■ Watts, D. J., and Strogatz, S. H. Collective dynamics of 'small-world' networks. *Nature, 393,* 440–442 (1998).

Subsequent work on the beta model, and even simpler related models, includes
◆ Barthelemy, M., and Amaral, L. A. N. Small-world networks: Evidence for a crossover picture. *Physical Review Letters, 82,* 3180–3183 (1999).
◆◆ Monasson, R. Diffusion, localization and dispersion relations on 'small-world' lattices. *European Physical Journal B,* 12(4), 555–567 (1999).
◆ Newman, M. E. J., and Watts, D. J. Scaling and percolation in the small-world network model. *Physical Review E, 60,* 7332–7342 (1999).
◆ Newman, M. E. J., and Watts, D. J. Renormalization group analysis of the small-world network model. *Physics Letters A, 263,* 341–346 (1999).
◆ Newman, M. E. J., Moore, C., and Watts, D. J. Mean-field solution of the small-world network model. *Physical Review Letters, 84,* 3201–3204 (2000).

Much of this early work is reviewed in
◆ Newman, M. E. J. Models of the small world. *Journal of Statistical Physics, 101,* 819–841 (2000).

The Real World

In addition to exploring the theoretical properties of small-world networks, researchers have identified a wide variety of empirical cases. Some examples are
■ Adamic, L. A. The small world web. In *Lecture Notes in Computer Science,* 1696, *Proceedings of the European Conference in Digital Libraries (ECDL) '99 Conference* (Springer, Berlin, 1999), pp. 443–454.

- Davis, G. F., Yoo, M., and Baker, W. E. The small world of corporate elite (working paper, University of Michigan Business School, 2002).
- Ferrer i Cancho, R., Janssen, C., and Solé, R. V. Topology of technology graphs: Small world patterns in electronic circuits. *Physical Review E*, 64, 046119 (2001).
- Kogut, B., and Walker, G. The small world of Germany and the durability of national networks. *American Sociological Review*, 66(3), 317–335 (2001).
- Sporns, O., Tononi, G., and Edelman, G. M. Theoretical neuroanatomy: Relating anatomical and functional connectivity in graphs and cortical connection matrices. *Cerebral Cortex*, 10, 127–141 (2000).
- Wagner, A., and Fell, D. The small world inside large metabolic networks. *Proceedings of the Royal Society of London, Series B*, 268, 1803–1810 (2001).

CHAPTER FOUR: BEYOND THE SMALL WORLD

Scale-Free Networks

A highly accessible account of the science of networks that focuses on the development and importance of scale-free networks is

- Barabási, A. L. *Linked: The New Science of Networks* (Perseus Press, Cambridge, MA, 2002).

Some more mathematical treatments of random networks with non-Poisson degree distributions (including power-law distributions) can be found in

♦♦ Aiello, W., Chung, F., and Lu, L. A random graph model for massive graphs. In *Proceedings of the 32nd Annual ACM Symposium on the Theory of Computing* (Association for Computing Machinery, New York, 2000), pp. 171–180.

♦♦ Molloy, M., and Reed, B. A critical point for random graphs with a given degree sequence. *Random Structures and Algorithms*, 6, 161–179 (1995).

♦♦ Molloy, M., and Reed, B. The size of the giant component of a random graph with a given degree sequence. *Combinatorics, Probability, and Computing*, 7, 295–305 (1998).

♦♦ Newman, M. E. J., Strogatz, S. H., and Watts, D. J. Random graphs with arbitrary degree distributions and their applications. *Physical Review E,* 64, 026118 (2001).

The Rich Get Richer

The paper by László Barabási and Réka Albert that introduced the idea of scale-free networks, and proposed a preferential attachment model of network growth to account for them, is
- Barabási, A., and Albert, R. Emergence of scaling in random networks. *Science,* 286, 509–512 (1999).

Since Barabási and Albert's original paper, a great many papers have been written on the subject of scale-free networks. Many of the references and relevant results are summarized in
- Albert, R., and Barabási, A. L. Statistical mechanics of complex networks. *Review of Modern Physics,* 74, 47–97 (2002).

Oddly enough, the original discovery of scale-free networks predates László Barabási and Réka Albert's paper by over thirty years. The empirical observation that networks can have a power-law degree distribution was first presented by Derek de Solla Price in
- Price, D. J. de Solla. Networks of scientific papers. *Science,* 149, 510–515 (1965).

Eleven years later, Price proposed a mathematical model that was, in essence, identical to that of Barabási and Albert's. Given how popular the notion of scale-free networks has become, one might wonder why no one picked up on it at the time. Perhaps the title (and the journal) had something to do with it.
- Price, D. J. de Solla. A general theory of bibliometrics and other cumulative advantage processes. *Journal of the American Society of Information Science,* 27, 292–306 (1976).

Continuing on an historical note, Zipf's law was first proposed in
- Zipf, G. K. *Human Behavior and the Principle of Least Effort* (Addison-Wesley, Cambridge, MA, 1949).

And Herbert Simon first presented the idea of preferential, random growth as an explanation of power-law size distributions like Zipf's law in

◆ Simon, H. A. On a class of skew distribution functions. *Biometrika*, 42, 425–440 (1955).

The article was reprinted two decades later, along with a significant amount of subsequent and related work, in
◆ Ijiri, Y., and Simon, H. A. *Skew Distributions and the Sizes of Business Firms* (Elsevier/North-Holland, New York, 1977).

Finally, the notion of the Matthew effect in the context of scientific prestige was introduced by Robert K. Merton in
• Merton, R. K. The Matthew effect in science. *Science*, 159, 56–63 (1968).

Getting Rich Can Be Hard

Empirical evidence (mostly) for and (occasionally) against the prevalence of scale-free networks has appeared in a number of places. Some of the more interesting papers are as follows:

■ Amaral, L. A. N., Scala, A., Barthelemy, M., and Stanley, H. E. Classes of behavior of small-world networks. *Proceedings of the National Academy of Sciences*, 97, 11149–11152 (2000).
■ Adamic, L. A., and Huberman, B. A. Power-law distribution of the World Wide Web. *Science*, 287, 2115a (2000).
■ Barabási, A. L., Albert, R., Jeong, H., and Bianconi, G. Power-law distribution of the World Wide Web, *Science*, 287, 2115b (2000).
■ Faloutsos, M., Faloutsos, P., and Faloutsos, C. On power-law relationships of the Internet topology. *Computer Communication Review*, 29, 251–262 (1999).
■ Liljeros, F., Edling, C. R., Amaral, L. A. N., Stanley, H. E., and Aberg, Y. The web of human sexual contacts. *Nature*, 411, 907–908 (2001).
◆ Rapoport, A. Mathematical models of social interaction. In Luce, R. D., Bush, R. R., and Galanter, E. (eds.), *Handbook of Mathematical Psychology*, Vol. 2 (Wiley, New York, 1963), pp. 493–579.
■ Redner, S. How popular is your paper? An empirical study of the citation distribution. *Europhysics Journal B*, 4, 131–134 (1998).

Reintroducing Group Structure

The paper Harrison presented that triggered our work on affiliation networks is

- White, H. C. What is the center of the small world? (paper presented at American Association for the Advancement of Science annual symposium, Washington, D.C., February 17–22, 2000).

And two classic references dealing with the importance of groups to the structure of social networks are
- Nadel, F. S. *Theory of Social Structure* (Free Press, Glencoe, IL, 1957).
- Breiger, R. L. The duality of persons and groups. *Social Forces, 53,* 181–190 (1974).

Affiliation Networks

A good basic reference for affiliation networks is
- Wasserman, S., and Faust, K. *Social Network Analysis: Methods and Applications* (Cambridge University Press, Cambridge, 1994).

Directors and Scientists

Jerry Davis's work on interlocking boards of corporate directors is presented in
- Davis, G. F. The significance of board interlocks for corporate governance. *Corporate Governance,* 4(3), 154–159 (1996).
- Davis, G. F., and Greve, H. R. Corporate elite networks and governance changes in the 1980s. *American Journal of Sociology,* 103(1), 1–37 (1997).

And Mark Newman's data on collaboration networks of scientists is
- Newman, M. E. J. The structure of scientific collaboration networks. *Proceedings of the National Academy of Sciences,* 98, 404–409 (2001).

Or in (slightly exhausting) detail:
- Newman, M. E. J. Scientific collaboration networks: I. Network construction and fundamental results. *Physical Review E,* 64, 016131 (2001).
- Newman, M. E. J. Scientific collaboration networks: II. Shortest paths, weighted networks, and centrality. *Physical Review E,* 64, 016132 (2001).

Complications

The mathematical machinery used to analyze the affiliation networks is described in

◆◆ Newman, M. E. J., Strogatz, S. H., and Watts, D. J. Random graphs with arbitrary degree distributions and their applications. *Physical Review E,* 64, 026118 (2001).

And a slightly more readable version is

◆ Newman, M. E. J., Watts, D. J., and Strogatz, S. H. Random graph models of social networks. *Proceedings of the National Academy of Sciences,* 99, 2566–2572 (2002).

CHAPTER FIVE: SEARCH IN NETWORKS

A compendium of Milgram's research over his entire, fascinating career is

● Milgram, S. *The Individual in a Social World: Essays and Experiments,* 2d ed. (McGraw-Hill, New York, 1992).

A detailed description of his obedience experiments is in

● Milgram, S. *Obedience to Authority: An Experimental View* (Harper & Row, New York, 1974).

So What Did Milgram Really Show?

Judith Kleinfeld's paper that exams the history and empirical validity of the small-world problem is

● Kleinfeld, J. S. The small world problem. *Society,* 39(2), 61–66 (2002).

The most significant follow-up study to the original experiment is one that Milgram conducted with another student, Charles Korte, in which they attempted to connect a white population of senders in Los Angeles with white and African-American targets in New York:

● Korte, C., and Milgram, S. Acquaintance networks between racial groups—application of the small world method. *Journal of Personality and Social Psychology,* 15(2), 101 (1970).

Is Six a Big or a Small Number?

The problem of Erdős numbers has been studied extensively by the mathematician Jerry Grossman, who maintains a Web page on the subject at http://www.oakland.edu/~grossman/erdoshp.html. An early summary of his work is

■ Grossman, J. W., and Ion, P. D. F. On a portion of the well-known collaboration graph. *Congressus Numerantium,* 108, 129–131 (1995).

Some more recent work on Erdős numbers is by
- Batagelj, V., and Mrvar, A. Some analyses of Erdős collaboration graph. *Social Networks,* 22(2), 173–186 (2000).

Other evidence that small-world networks can make problem-solving harder, rather than easier, is presented in
- Walsh, T. Search in a small world. In *Proceedings of the 16th International Joint Conference on Artificial Intelligence* (Morgan Kaufmann, San Francisco 1999), pp. 1172–1177.

The Small-World Search Problem

Jon Kleinberg's groundbreaking paper, which pointed out and then solved the small-world search problem, is available in two versions, long and short, respectively:
- Kleinberg, J. The small-world phenomenon: An algorithmic perspective. In *Proceedings of the 32nd Annual ACM Symposium on Theory of Computing* (Association of Computing Machinery, New York, 2000), pp. 163–170.
- Kleinberg, J. Navigation in a small world. *Nature,* 406, 845 (2000).

Kleinberg later used a similar approach to study the spread of information in networks via what computer scientists call a *gossip protocol:*
- Kleinberg, J. Small-world phenomena and the dynamics of information. In Dietterich, T. G., Becker, S., and Ghahramani, Z. (eds.), *Advances in Neural Information Processing Systems* (NIPS), 14 (MIT Press, Cambridge, MA, 2002).

Sociology Strikes Back

The paper that resulted from my collaboration with Mark Newman and Peter Dodds, incorporating notions of social identity and social distance into the small-world search problem, is
- Watts, D. J., Dodds, P. S., and Newman, M. E. J. Identity and search in social networks. *Science,* 296, 1302–1305 (2002).

The findings of the so-called reverse small-world experiment that corroborated some of our theoretical predictions were published in

- Killworth, P. D., and Bernard, H. R. The reverse small world experiment. *Social Networks,* 1, 159–192 (1978).
- Bernard, H. R., Killworth, P. D., Evans, M. J., McCarty, C., and Shelly, G. A. Studying relations cross-culturally. *Ethnology,* 27(2), 155–179 (1988).

Search in Peer-to-Peer Networks

A discussion of the problems facing peer-to-peer networks like Gnutella is

- Ritter, J. P. Why Gnutella can't scale. No really (working paper, available on-line at http://www.darkridge.com/~jpr5/doc/gnutella.html, 2000).

Two search algorithms that take advantage of Gnutella's apparent scale-free character are presented in

- ◆ Adamic, L. A., Lukose, R. M., Puniyani, A. R., and Huberman, B. A. Search in power-law networks. *Physical Review E,* 64, 046135 (2001).
- ◆ Kim, B. J., Yoon, C. N., Han, S. K., and Jeong, H. Path finding strategies in scale-free networks. *Physical Review E,* 65, 027103 (2002).

And a discussion of the problems associated with creating easily searchable distributed databases in the context of a multinational consulting firm is

- Mannville, B. Complex adaptive knowledge management: A case study from McKinsey and Company. In Clippinger, J. H. (ed.), *The Biology of Business: Decoding the Natural Laws of the Enterprise* (Jossey-Bass, San Francisco, 1999), chapter 5.

Some other approaches that have been proposed specifically for the purpose of finding information on the World Wide Web are dealt with in the following:

- Brin, S., and Page, L. The anatomy of a large-scale hypertextual web search engine. *Computer Networks,* 30, 107–117 (1998).
- ◆ Gibson, D., Kleinberg, J., and Raghavan, P. Inferring Web communities from link topology. In *Proceedings of the 9th ACM Conference on Hypertext and Hypermedia* (Association for Computing Machinery, Networks 1998), pp. 225–234.
- ◆ Kleinberg, J. Authoritative sources in a hyperlinked environment. *Journal of the ACM,* 46, 604–632 (1999).

- Lawrence, S., and Giles, C. L. Accessibility of information on the web. *Nature,* 400, 107–109 (1999).

CHAPTER SIX: EPIDEMICS AND FAILURES

The Hot Zone

Richard Preston's gripping account of the Ebola outbreak in Reston, Virginia, along with a brief history of Ebola, can be found in his book
- Preston, R. *The Hot Zone* (Random House, New York, 1994).

Additional Ebola facts were derived from
- Harden, B. Dr. Matthew's passion. *New York Times Magazine,* February 18, 2001, pp. 24–62.

Viruses in the Internet

The account of the e-mail sent by Claire Swire was taken from a *New York Times* article:
- Lyall, S. Return to sender, please. *New York Times,* December 24, 2000, *Week in Review,* p. 2.

A chronology of the Melissa virus can be found at http://www.cert.org/advisories/CA-1999-04.html.

Records of all registered computer viruses and their histories, including time of first detection, known number of computers infected, and first release of antiviral software, can be found at the Virus Bulletin: http://www.virusbtn.com/. Virus alerts, along with other Internet-related security information, are published and maintained by CERT, based in the Software Engineering Institute at Carnegie Mellon University in Pittsburgh. Their Web site is http://www.cert.org/.

A discussion of the relationship between epidemiology and computer viruses in the age of the Internet is contained in the following:
- Kephart, J. O., White, S. R., and Chess, D. M. Computer viruses and epidemiology. *IEEE Spectrum,* 30(5), 20–26 (1993).

- Kephart, J. O., Sorkin, G. B., Chess, D. M., and White, S. R. Fighting computer viruses. *Scientific American*, 277(5), 56–61 (1997).

The Mathematics of Epidemics

The classic papers of Kermack and McKendrick, on which most of modern mathematical epidemiology is based, are
◆◆ Kermack, W. O., and McKendrick, A. G. A contribution to the mathematical theory of epidemics. *Proceedings of the Royal Society of London, Series A*, 115, 700–721 (1927).
◆◆ Kermack, W. O., and McKendrick, A. G. Contributions to the mathematical theory of epidemics. II. The problem of endemicity. *Proceedings of the Royal Society of London, Series A*, 138, 55–83 (1932).
◆◆ Kermack, W. O., and McKendrick, A. G. Contributions to the mathematical theory of epidemics. III. Further studies of the problem of endemicity. *Proceedings of the Royal Society of London, Series A*, 141, 94–122 (1933).

The standard text of mathematical epidemiology, and one that treats the SIR model in considerable detail, is
◆ Bailey, N. T. J. *The Mathematical Theory of Infectious Diseases and Its Applications* (Hafner Press, New York, 1975).

Other good references are
◆ Bartholomew, D. J. *Stochastic Models for Social Processes* (Wiley, New York, 1967).
◆ Anderson, R. M., and May, R. M. *Infectious Diseases of Humans* (Oxford University Press, Oxford, 1991).
◆ Murray, J. D. *Mathematical Biology*, 2d ed. (Springer, Heidelberg, 1993).

A small but excellent collection of papers dealing with the spread of infectious diseases on networks is
◆◆ Ball, F., Mollison, D., and Scalia-Tomba, G. Epidemics with two levels of mixing. *Annals of Applied Probability*, 7(1), 46–89 (1997).
- Hess, G. Disease in metapopulation models: Implications for conservation. *Ecology*, 77, 1617–1632 (1996).
- Kareiva, P. Population dynamics in spatially complex environments:

Theory and data. *Philosophical Transactions of the Royal Society of London, Series B,* 330, 175–190 (1988).

♦ Kretschmar, M., and Morris, M. Measures of concurrency in networks and the spread of infectious disease. *Mathematical Biosciences,* 133, 165–195 (1996).

■ Longini, I. M., Jr. A mathematical model for predicting the geographic spread of new infectious agents. *Mathematical Biosciences,* 90, 367–383 (1988).

♦ Sattenspiel, L., and Simon, C. P. The spread and persistence of infectious diseases in structured populations. *Mathematical Biosciences,* 90, 341–366 (1988).

Epidemics in a Small World

The most complete account of the early work on disease spreading on small-world networks is chapter 6 of

■ Watts, D. J. *Small Worlds: The Dynamics of Networks between Order and Randomness* (Princeton University Press, Princeton, NJ, 1999).

A number of subsequent results about epidemics on networks have been published in the following papers:

♦ Boots, M., and A. Sasaki. "Small worlds" and the evolution of virulence: Infection occurs locally and at a distance. *Proceedings of the Royal Society of London, Series B,* 266, 1933–1938 (1999).

♦ Keeling, M. J. The effects of local spatial structure on epidemiological invasions. *Proceedings of the Royal Society of London, Series B,* 266, 859–867 (1999).

♦ Kuperman, M., and Abramson, G. Small world effect in an epidemiological model. *Physical Review Letters,* 86, 2909–2912 (2001).

Some excellent works on the foot-and-mouth epidemic, and a fine example of mathematical modeling contributing to policy decisions, are

■ Ferguson, N. M., Donnelly, C. A., and Anderson, R. M. The foot-and-mouth epidemic in Great Britain: Pattern of spread and impact of interventions. *Science,* 292, 1155–1160 (2001).

■ Ferguson, N. M., Donnelly, C. A., and Anderson, R. M. Transmission intensity and impact of control policies on the foot and mouth epidemic in Great Britain. *Nature,* 413, 542–548 (2001).

■ Keeling, M. J., Woolhouse, M. E. J., Shaw, D. J., Matthews, L., Chase-Topping, M., Haydon, D. T., Cornell, S. J., Kappey, J., Wilesmith, J., and Grenfell, B. T. Dynamics of the 2001 UK foot and mouth epidemic: Stochastic dispersal in a heterogeneous landscape. *Science*, 294, 813–817 (2001).

The discovery that diseases spreading in scale-free networks don't exhibit an epidemic threshold is reported in
◆ Pastor-Satorras, R., and Vespignani, A. Epidemic spreading in scale-free networks. *Physical Review Letters*, 86, 3200–3203 (2001).

Pastor-Satorras and Vespignani have continued to work on disease spreading in scale-free networks. Their findings are summarized in
◆ Pastor-Satorras, R., and Vespignani, A. Epidemics and immunization in scale-free networks. In Bornholdt, S., and Schuster, H. G. (eds.), *Handbook of Graphs and Networks: From the Genome to the Internet* (Wiley-VCH, Berlin, 2002).

Some empirical support for their assumption of a scale-free e-mail network is reported in
◆ Ebel, H., Mielsch, L. I., and Bornholdt, S. Scale-free topology of e-mail networks. *Physical Review E*, 66, 035103 (2002).

The public health impact of needle exchange programs is analyzed in a report for the *Journal of the American Medical Association* by the Center for AIDS Prevention Studies, University of California, San Francisco, found at http://www.ama-assn.org/special/hiv/preventn/prevent3.htm.

An earlier report prepared for the Centers for Disease Control and Prevention can be found at http://www.caps.ucsf.edu/capsweb/publications/needlereport.html.

Percolation Models of Disease

The best introduction to the subject of percolation is the following (which even manages to be funny in places):
◆ Stauffer, D., and Aharony, A. *Introduction to Percolation Theory* (Taylor and Francis, London, 1992).

The details of my work with Mark Newman on a site percolation approach to disease spreading on small-world network are given in

♦ Newman, M. E. J., and Watts, D. J. Scaling and percolation in the small-world network model. *Physical Review E*, 60, 7332–7342 (1999).

Networks, Viruses, and Microsoft

Mark Newman and Cris Moore's work on site and bond percolation is described in the following

♦ Moore, C., and Newman, M. E. J. Epidemics and percolation in small-world networks. *Physical Review E*, 61, 5678–5682 (2000).

♦♦ Moore, C., and Newman, M. E. J. Exact solution of site and bond percolation on small-world networks. *Physical Review E*, 62, 7059–7064 (2000).

Failures and Robustness

The original paper using percolation ideas to quantify network robustness is

■ Albert, R., Jeong, H., and Barabási, A. L. Attack and error tolerance of complex networks. *Nature*, 406, 378–382 (2000).

Soon after, a series of papers examined the topic in greater detail. They are

♦ Callaway, D. S., Newman, M. E. J., Strogatz, S. H., and Watts, D. J. Network robustness and fragility: Percolation on random graphs. *Physical Review Letters*, 85, 5468–5471 (2000).

♦ Cohen, R., Erez, K., ben-Avraham, D., and Havlin, S. Resilience of the Internet to random breakdowns. *Physical Review Letters*, 85, 4626–4628 (2000).

♦ Cohen, R., Erez, K., ben-Avraham, D., and Havlin, S. Breakdown of the Internet under intentional attack. *Physical Review Letters*, 86, 3682–3685 (2001).

CHAPTER SEVEN: DECISIONS, DELUSIONS, AND THE MADNESS OF CROWDS

Tulip Economics

Charles Mackay's classic account of mania, financial and otherwise, has been reprinted many times. A relatively recent version is

- Mackay, C. *Extraordinary Popular Delusions and the Madness of Crowds* (Harmony Books, New York, 1980).

Some other recent treatises on the same subject are
- Kindleberger, C. P. *Manias, Panics, and Crashes: A History of Financial Crises,* 4th ed. (Wiley, New York, 2000).
- Shiller, R. J. Irrational Exuberance (Princeton University Press, Princeton, NJ, 2000).

Fear, Greed, and Rationality

Adam Smith's discussion of rationally optimizing agents, including his reference to the invisible hand, is in
- Smith, A. *The Wealth of Nations,* Vol. 1, Book 4 (University of Chicago Press, Chicago, 1976), chapter 2, p. 477.

The paradox of the efficient market hypothesis is described in
- Chancellor, E. *Devil Take the Hindmost: A History of Financial Speculation* (Farrar, Straus and Giroux, New York, 1999).

And some recent work on constructing a more realistic vision of both investor and financial-market behavior, in which dynamics is a crucial ingredient, is
- Farmer, J. D. Market force, ecology, and evolution. *Industrial and Corporate Change,* forthcoming (2002).
- Farmer, J. D., and Joshi, S. The price dynamics of common trading strategies. *Journal of Economic Behavior and Organization,* 49(2), 149–171 (2002).
- Farmer, J. D., and Lo, A. Frontiers of finance: Evolution and efficient markets. *Proceedings of the National Academy of Sciences,* 96, 9991–9992 (1999).

Collective Decisions

The technical paper by Natalie Glance and Bernardo Huberman, describing the diner's dilemma and the conditions under which it can be resolved, is
- Glance, N. S., and Huberman, B. A. The outbreak of cooperation. *Journal of Mathematical Sociology,* 17(4), 281–302 (1993).

A more accessible description of the same results is
- Glance, N. S., and Huberman, B. A. The dynamics of social dilemmas. *Scientific American,* 270(3), 76–81 (1994).

The literature on the evolution of cooperation is gigantic and spans several disciplines—evolutionary biology, economics, political science, and sociology in particular. It is impossible even to provide a representative list of publications, but some important contributions are
- Axelrod, R. *The Evolution of Cooperation* (Basic Books, New York, 1984).
- Axelrod, R., and Dion, D. The further evolution of cooperation. *Science,* 242, 1385–1390 (1988).
- Boorman, S. A., and Levitt, P. R. *The Genetics of Altruism* (Academic Press, New York, 1980).
- Boyd, R. S., and Richerson, P. J. The evolution of reciprocity in sizable groups. *Journal of Theoretical Biology,* 132, 337–356 (1988).
- Hardin, G. The tragedy of the commons. *Science,* 162, 1243–1248 (1968).
- Huberman, B. A., and Lukose, R. M. Social dilemmas and Internet congestion. *Science,* 277, 535–537 (1997).
- Nowak, M. A., and May, R. M. Evolutionary games and spatial chaos. *Nature,* 359, 826–829 (1992).
- Olson, M. *The Logic of Collective Action: Public Goods and the Theory of Groups* (Harvard University Press, Cambridge, MA, 1965).
- Ostrom, E., Burger, J., Field, C. B., Norgaard, R. B., and Policansky, D. Revisiting the commons: Local lessons, global challenges. *Science,* 284, 278–282 (1999).

Information Cascades

The literature on information cascades is, again, multidisciplinary and eclectic. Some examples are
- Aguirre, B. E., Quarantelli, E. L., and Mendoza, J. L. The collective behavior of fads: The characteristics, effects, and career of streaking. *American Sociological Review,* 53, 569–584 (1988).
- Banerjee, A. V. A simple model of herd behavior. *Quarterly Journal of Economics,* 107, 797–817 (1992).
- Bikhchandani, S., Hirshleifer, D., and Welch, I. A theory of fads, fash-

ion, custom and cultural change as informational cascades. *Journal of Political Economy,* 100(5), 992–1026 (1992).

■ Lohmann, S. The dynamics of informational cascades: The Monday demonstrations in Leipzig, East Germany, 1989–91. *World Politics,* 47, 42–101 (1994).

Information Externalities

A description of Asch's original experiment is given in

● Asch, S. E. Effects of group pressure upon the modification and distortion of judgments. In Cartwright, D., and Zander, A. (eds.), *Group Dynamics: Research and Theory* (Row, Peterson, Evanston, IL, 1953), pp. 151–162.

Herbert Simon's theory of bounded rationality is espoused in

■ Simon, H. A., Egidi, M., and Marris, R. L. *Economics, Bounded Rationality and the Cognitive Revolution* (Edward Elgar, Brookfield, VT, 1992).

Coercive Externalities

The spread of crime via a network of peer pressure relationships is considered (in a theoretical manner) in

■ Glaeser, E. L., Sacerdote, B., and Schheinkman, J. A. Crime and social interactions. *Quarterly Journal of Economics,* 111, 507–548 (1996).

The paper introducing the concept of a spiral of silence in voting behavior is

● Noelle-Neumann, E. Turbulences in the climate of opinion: Methodological applications of the spiral of silence theory. *Public Opinion Quarterly,* 41(2), 143–158 (1977).

Market Externalities

The principal exponent of what is now called lock-in via increasing returns is the economist Brian Arthur. His groundbreaking paper (which took him many years to find a journal that would publish it) is

◆ Arthur, W. B. Competing technologies, increasing returns, and lock-in by historical events. *Economic Journal,* 99(394), 116–131 (1989).

Another, slightly different, approach to the topic of increasing returns is espoused in
♦ Romer, P. Increasing returns and long-run growth. *Journal of Political Economy*, 94(5), 1002–1034 (1986).

Although the authors do not link it to the subject of network externalities, the importance of complementarities is emphasized in
♦ Milgrom, P., and Roberts, J. The economics of modern manufacturing: Technology, strategy, and organization. *American Economic Review*, 80(3), 511–528 (1990).

Meanwhile, the prevailing economics approach to so-called network externalities is described in
■ Economides, N. The economics of networks. *International Journal of Industrial Organization*, 16(4), 673–699 (1996).

Coordination Externalities

Although neither of these papers uses the term, the relevance of externalities to decisions about cooperation is evident in
♦ Glance, N. S., and Huberman, B. A. The outbreak of cooperation. *Journal of Mathematical Sociology*, 17(4), 281–302 (1993).
■ Kim, H., and Bearman, P. The structure and dynamics of movement participation. *American Sociological Review*, 62(1), 70–94 (1997).

Social Decision Making

The report on body piercing is
● Harden, B. Coming to grips with the enduring appeal of body piercing. *New York Times*, February 12, 2002, p. A16.

CHAPTER EIGHT: THRESHOLDS, CASCADES, AND PREDICTABILITY

Threshold Models of Decisions

The earliest use of threshold models as a way of understanding collective decision making is probably
■ Schelling, T. C. A study of binary choices with externalities. *Journal of Conflict Resolution*, 17(3), 381–428 (1973).

Another early classic is
- Granovetter, M. Threshold models of collective behavior. *American Journal of Sociology,* 83(6), 1420–1443 (1978).

The actual derivation of a threshold model depends on what kind of decision an individual is making, and what sort of decision externalities pertain. Some examples of quite different derivations, all of which effectively generate threshold rules are
- Arthur, W. B., and Lane, D. A. Information contagion. *Structural Change and Economic Dynamics,* 4(1), 81–103 (1993).
- Boorman, S. A., and Levitt, P. R. *The Genetics of Altruism* (Academic Press, New York, 1980).
- Durlauf, S. N. A framework for the study of individual behavior and social interactions. *Sociological Methodology,* 31, 47–87 (2001).
- Glance, N. S., and Huberman, B. A. The outbreak of cooperation. *Journal of Mathematical Sociology,* 17(4), 281–302 (1993).
- Morris, S. N. Contagion. *Review of Economic Studies,* 67, 57–78 (2000).

Cascades in Social Networks

Some information about *Ithaca hours* can be found in
- Glover, P. Grassroots economics. *In Context,* 41, 30 (1995).
- Morse, M. Dollars or sense. *Utne Reader,* 99 (September–October 1999).

Everett Rogers' classic reference on the diffusion of innovations, in which much of the still-current terminology was introduced, was first published in 1962. It is now in its fourth edition:
- Rogers, E. *The Diffusion of Innovations,* 4th ed. (Free Press, New York, 1995).

A valiant attempt to combine Rogers' ideas with concepts from social network analysis is by one of Rogers' students, Thomas Valente:
- Valente, T. W. *Network Models of the Diffusion of Innovations* (Hampton Press, Cresskill, NJ, 1995).

Cascades and Percolation

The paper summarizing the threshold model approach to information cascades on networks is

◆ Watts, D. J. A simple model of global cascades on random networks. *Proceedings of the National Academy of Sciences*, 99, 5766–5771 (2002).

Phase Transitions and Cascades

Malcolm Gladwell's engaging discussion of social contagion is

● Gladwell, M. *The Tipping Point: How Little Things Can Make a Big Difference* (Little, Brown, New York, 2000).

Crossing the Chasm

Geoffrey Moore's descriptive account of the "chasm" between early adopters and the early to late majority is given in

● Moore, G. A. *Crossing the Chasm: Marketing and Selling High-Tech Products to Mainstream Customers* (Harper Business, New York, 1999).

A Nonlinear View of History

The distinction between quality and success is made apparent in Art De Vany's study of the movie industry:

■ De Vany, A., and Lee, C. Quality signals in information cascades and the dynamics of motion picture box office revenues: A computational model. *Journal of Economic Dynamics and Control*, 25, 593–614 (2001).
◆ De Vany, A. S., and Walls, W. D. Bose-Einstein dynamics and adaptive contracting in the motion picture industry. *Economic Journal*, 106, 1493–1514 (1996).

Robustness Revisited

The related notions of normal accidents and robust-yet-fragile systems are presented in two very different works:

● Perrow, C. *Normal Accidents: Living with High-Risk Technologies* (Basic Books, New York, 1984).
◆ Carlson, J. M., and Doyle, J. Highly optimized tolerance: A mechanism for power laws in designed systems. *Physical Review E*, 60(2), 1412–1427 (1999).

CHAPTER NINE: INNOVATION, ADAPTATION, AND RECOVERY

The Toyota-Aisin Crisis

The account of the Toyota-Aisin crisis on which my description is based is
• Nishiguchi, T., and Beaudet, A. Fractal design: Self-organizing links in supply chain management. In Von Krogh, G., Nonaka, I., and Nishiguchi, T. (eds.), *Knowledge Creation: A Source of Value* (Macmillan, London, 2000).

Another paper about the remarkable Toyota group that guided our thinking on innovation is
• Ward, A., Liker, J. K., Cristiano, J. J., and Sobek, D. K. The second Toyota paradox: How delaying decisions can make better cars faster. *Sloan Management Review*, 36(3), 43–51 (1995).

Markets and Hierarchies

The original text—and still one of the greatest—on industrial organization is
• Smith, A. *The Wealth of Nations* (University of Chicago Press, Chicago, 1976).

A precursor to Coase's theory of transaction costs was Frank Knight's claim that firms exist to reduce uncertainty:
▪ Knight, F. H. *Risk, Uncertainty, and Profit* (London School of Economics and Political Science, London, 1933).

And Ronald Coase's original argument of transaction costs as the basis for the firm is explicated in
• Coase, R. The nature of the firm. *Economica*, n.s., 4 (November 1937).

Several decades later, Coase is still trying to get his ideas accepted by mainstream economics. His latest attempt is
• Coase, R. *The Nature of the Firm* (Oxford University Press, Oxford, 1991).

The chief proponent of the hierarchical structure of firms is Oliver Williamson, whose views are expressed comprehensively in

- Williamson, O. E. *Markets and Hierarchies* (Free Press, New York, 1975).

A shorter version is
- Williamson, O. E. Transaction cost economics and organization theory. In Smelser, N. J., and Swedberg, R. (eds.), *The Handbook of Economic Sociology* (Princeton University Press, Princeton, NJ, 1994), pp. 77–107.

The superiority of the hierarchy has been extensively formalized and developed in recent years by a small group of economists, led by Roy Radner. Some of the principal works of this literature are
- Bolton, P., and Dewatripont, M. The firm as a communication network. *Quarterly Journal of Economics*, 109(4), 809–839 (1994).
- Radner, R. The organization of decentralized information processing. *Econometrica*, 61(5), 1109–1146 (1993).
- Radner, R. Bounded rationality, indeterminacy, and the theory of the firm. *Economic Journal*, 106, 1360–1373 (1996).
- Van Zandt, T. Decentralized information processing in the theory of organizations. In Sertel, M. (ed.), *Contemporary Economic Issues*, vol. 4: *Economic Design and Behavior* p. 125–160 (Macmillan, London, 1999), chapter 7.

Industrial Divides

Michael Piore and Chuck Sabel's groundbreaking book on the changing nature of the global economy is
- Piore, M. J., and Sabel, C. F. *The Second Industrial Divide: Possibilities for Prosperity* (Basic Books, New York, 1984).

Ambiguity

The paper describing Honda's system for identifying problems in their manufacturing plants is
- MacDuffie, J. P. The road to "root cause": Shop-floor problem-solving at three auto assembly plants. *Management Science*, 43, 4 (1997).

A variety of approaches to the theory of the internal architecture of the firm have been taken within the economics, sociology, and business communities. It is a vast literature. An eclectic collection of readings, with no claim to being exhaustive or even representative, is as follows:

• Chandler, A. D. *The Visible Hand: The Managerial Revolution in American Business* (Belknap Press of Harvard University Press, Cambridge, MA, 1977).

• Clippinger, J. (ed.). *The Biology of Business: Decoding the Natural Laws of the Enterprise* (Jossey-Bass, San Francisco, 1999).

♦ Fama, E. F. Agency problems and the theory of the firm. *Journal of Political Economy*, 88, 288–307 (1980).

■ Hart, O. *Firms, Contracts and Financial Structure* (Oxford University Press, New York, 1995).

■ March, J. G., and Simon, H. A. *Organizations* (Blackwell, Oxford, 1993).

■ Nelson, R. R., and Winter, S. G. *An Evolutionary Theory of Economic Change* (Belknap Press of Harvard University Press, Cambridge, MA, 1982).

■ Powell, W., and DiMaggio, P. (eds.). *The New Institutionalism in Organizational Analysis* (Chicago, University of Chicago Press, 1991).

♦ Sah, R. K., and Stiglitz, J. E. The architecture of economic systems: Hierarchies and polyarchies. *American Economic Review*, 76(4), 716–727 (1986).

The Third Way

An indication of Chuck's understanding of the problem at the time we started working together is given in the following:

• Helper, S., MacDuffie, J. P., and Sabel, C. F. Pragmatic collaborations: Advancing knowledge while controlling opportunism. *Industrial and Corporate Change*, 9(3), 443–488 (2000).

• Sabel, C. F. Diversity, not specialization: The ties that bind the (new) industrial district. In Quadrio Curzio, A., and Fortis, M. (eds.), *Complexity and Industrial Clusters: Dynamics and Models in Theory and Practice* (Physica-Verlag, Heidelberg, 2002).

Coping with Ambiguity

Perhaps the clearest exposition of the conundrum faced by firms in ambiguous environments, and their need to be both adapted and adaptable, is that by David Stark in his work on *heterarchies:*

• Stark, D. C. Recombinant property in East European capitalism. *American Journal of Sociology*, 101(4), 993–1027 (1996).

• Stark, D. C. Heterarchy: Distributing authority and organizing diversity. In Clippinger, J. H. (ed.), *The Biology of Business: Decoding the Natural Laws of the Enterprise* (Jossey-Bass, San Francisco, 1999), chapter 7.

• Stark, D. C., and Bruszt, L. *Postsocialist Pathways: Transforming Politics and Property in East Central Europe* (Cambridge University Press, Cambridge, 1998).

Multiscale Networks

The properties of team-based, multiscale, and core-periphery networks are outlined in

◆ Dodds, P. S., Watts, D. J., and Sabel, C. F. The structure of optimal redistribution networks. Institute for Social and Economic Research and Policy Working Paper, Columbia University (2002).

CHAPTER TEN: THE END OF THE BEGINNING

The use of New York City (or more precisely, the borough of Manhattan) as an example of a complex system was inspired by John Holland's opening chapter in

• Holland, J. H. *Hidden Order: How Adaptation Builds Complexity* (Perseus, Cambridge, MA, 1996).

September 11

An excellent and detailed account of the September 11 attacks on the World Trade Center, and the monumental recovery effort that followed its collapse, is given in

• Langewiesche, W. *American Ground: Unbuilding the World Trade Center* (North Point Press, New York, 2002).

The information about the police department's communications predicament was derived from

• Rashbaum, W. K. Police officers swiftly show inventiveness during crisis. *New York Times*, September 17, 2001, p. A7.

The story about Cantor Fitzgerald was related by one of the survivors—the director of marketing and communications—who spoke at a roundtable discussion of business leaders at Columbia University on December 5, 2001. The roundtable was organized by David Stark and John Kelly, directors of Columbia's Center for Organizational Innovation and Interactive Design Lab, respectively, and sponsored by Dr. Susan Gitelson.

The article by Paul Krugman considering the economic consequences of the September 11 attacks in the context of an already faltering economy is
- Krugman, P. Fear itself. *New York Times Magazine*, September 30, 2001, p. 36.

Lessons for a Connected Age

An engaging and insightful account of the 1997 Asian crisis is contained in
- Friedman, T. L. *The Lexus and the Olive Tree: Understanding Globalization* (Farrar, Straus and Giroux, New York, 1999).

Quite a lot of Harry Potter–related information can be found at http://www.insideharrypotter.com.

And finally, a brief but illuminating account of the problems that beset Long Term Capital Management in the fall of 1998, is
- MacKenzie, D. Fear in the markets. *London Review of Books*, 22(8), 31–32 (2000).

Bibliography

Adamic L. A. The small world web. In *Lecture Notes in Computer Science*, 1696, *Proceedings of the European Conference on Digital Libraries (ECDL) '99 Conference* (Springer, Berlin, 1999), pp. 443–454.

Adamic, L. A., and Huberman, B. A. Power-law distribution of the World Wide Web. *Science*, 287, 2115a (2000).

Adamic, L. A., Lukose, R. M., Puniyani, A. R., and Huberman, B. A. Search in power-law networks. *Physical Review E*, 64, 046135 (2001).

Aguirre, B. E., Quarantelli, E. L., and Mendoza, J. L. The collective behavior of fads: The characteristics, effects, and career of streaking. *American Sociological Review*, 53, 569–584 (1988).

Aho, A. V., Hopcroft, J. E., and Ullman, J. D. *Data Structures and Algorithms* (Addison Wesley, Reading, MA, 1983).

Ahuja, R. K., Magnanti, T. L., and Orlin, J. B. *Network Flows: Theory, Algorithms, and Applications* (Prentice Hall, Englewood Cliffs, NJ, 1993).

Aiello, W., Chung, F., and Lu, L. A random graph model for massive graphs. In *Proceedings of the 32nd Annual ACM Symposium on the Theory of Computing* (Association for Computing Machinery, New York, 2000), pp. 171–180.

Albert, R., and Barabási, A. L. Statistical mechanics of complex networks. *Review of Modern Physics*, 74, 47–97 (2002).

Albert, R., Jeong, H., and Barabási, A. L. Attack and error tolerance of complex networks. *Nature*, 406, 378–382 (2000).

Alon, N., and Spencer, J. H. *The Probabilistic Method* (Wiley-Interscience, New York, 1992).

Amaral, L. A. N., Scala, A., Barthelemy, M., and Stanley, H. E. Classes of behavior of small-world networks. *Proceedings of the National Academy of Sciences*, 97, 11149–11152 (2000).

Anderson, P. W. More is different. *Science,* 177, 393–396 (1972).

Anderson, R. M., and May, R. M. *Infectious Diseases of Humans* (Oxford University Press, Oxford, 1991).

Arthur, W. B. Competing technologies, increasing returns, and lock-in by historical events. *Economic Journal,* 99(394), 116–131 (1989).

Arthur, W. B., and Lane, D. A. Information contagion. *Structural Change and Economic Dynamics,* 4(1), 81–103 (1993).

Asavathiratham, C. *The Influence Model: A Tractable Representation for the Dynamics of Networked Markov Chains.* Ph.D. Dissertation, Department of Electrical Engineering and Computer Science, MIT (MIT, Cambridge, MA, 2000).

Asch, S. E. Effects of group pressure upon the modification and distortion of judgments. In Cartwright, D., and Zander, A. (eds.), *Group Dynamics: Research and Theory* (Row, Peterson, Evanston, IL, 1953), pp. 151–162.

Asimov, I. *The Caves of Steel* (Doubleday, Garden City, NY, 1954).

———. *The Naked Sun* (Doubleday, Garden City, NY, 1957).

Axelrod, R. *The Evolution of Cooperation* (Basic Books, New York, 1984).

Axelrod, R., and Dion, D. The further evolution of cooperation. *Science,* 242, 1385–1390 (1988).

Bailey, N. T. J. *The Mathematical Theory of Infectious Diseases and Its Applications* (Hafner Press, New York, 1975).

Ball, F., Mollison, D., and Scalia-Tomba, G. Epidemics with two levels of mixing. *Annals of Applied Probability,* 7(1), 46–89 (1997).

Banerjee, A. V. A simple model of herd behavior. *Quarterly Journal of Economics,* 107, 797–817 (1992).

Barabási, A., and Albert, R. Emergence of scaling in random networks. *Science,* 286, 509–512 (1999).

Barabási, A. L. *Linked: The New Science of Networks* (Perseus Press, Cambridge, MA, 2002).

Barabási, A. L., Albert, R., Jeong, H., and Bianconi, G. Power-law distribution of the World Wide Web. *Science,* 287, 2115b (2000).

Barthelemy, M., and Amaral, L. A. N. Small-world networks: Evidence for a crossover picture. *Physical Review Letters,* 82, 3180–3183 (1999).

Bartholomew, D. J. *Stochastic Models for Social Processes* (Wiley, New York, 1967).

Batagelj, V., and Mrvar, A. Some analyses of Erdős collaboration graph. *Social Networks*, 22(2), 173–186 (2000).

Bernard, H. R., Killworth, P. D., Evans, M. J., McCarty, C., and Shelly, G. A. Studying relations cross-culturally. *Ethnology*, 27(2), 155–179 (1988).

Bickhchandani, S., Hirshleifer, D., and Welch, I. A theory of fads, fashion, custom and cultural change as informational cascades. *Journal of Political Economy*, 100(5), 992–1026 (1992).

Bollobas, B. *Random Graphs*, 2nd ed. (Academic, New York, 2001).

Bolton, P., and Dewatripont, M. The firm as a communication network. *Quarterly Journal of Economics*, 109(4), 809–839 (1994).

Boorman, S. A., and Levitt, P. R. *The Genetics of Altruism* (Academic Press, New York, 1980).

Boorman, S. A., and White, H. C. Social structure from multiple networks. II. Role structures. *American Journal of Sociology*, 81(6), 1384–1446 (1976).

Boots, M., and A. Sasaki. "Small worlds" and the evolution of virulence: Infection occurs locally and at a distance. *Proceedings of the Royal Society of London, Series B*, 266, 1933–1938 (1999).

Boyd, R. S., and Richerson, P. J. The evolution of reciprocity in sizable groups. *Journal of Theoretical Biology*, 132, 337–356 (1988).

Breiger, R. L. The duality of persons and groups. *Social Forces*, 53, 181–190 (1974).

Brin, S., and Page, L. The anatomy of a large-scale hypertextual web search engine. *Computer Networks*, 30, 107–117 (1998).

Burt, R. S. *Structural Holes: The Social Structure of Competition* (Harvard University Press, Cambridge, MA, 1992).

Callaway, D. S., Newman, M. E. J., Strogatz, S. H., and Watts, D. J. Network robustness and fragility: Percolation on random graphs. *Physical Review Letters*, 85, 5468–5471 (2000).

Carlson, J. M., and Doyle, J. Highly optimized tolerance: A mechanism for power laws in designed systems. *Physical Review E*, 60(2), 1412–1427 (1999).

Casti, J. L. *Reality Rules I & II: Picturing the World in Mathematics: The Fundamentals, the Frontier* (Wiley-Interscience, New York, 1997).

Chancellor, E. *Devil Take the Hindmost: A History of Financial Speculation* (Farrar, Straus and Giroux, New York, 1999).

Chandler, A. D. *The Visible Hand: The Managerial Revolution in*

American Business (Belknap Press of Harvard University Press, Cambridge, MA, 1977).

Clippinger, J. (ed.) *The Biology of Business: Decoding the Natural Laws of the Enterprise* (Jossey-Bass, San Francisco, 1999).

Coase, R. The nature of the firm. *Economica,* n.s., 4 (November 1937).

———. *The Nature of the Firm* (Oxford University Press, Oxford 1991).

Cohen, R. Who really brought down Milosevic? *New York Times Magazine,* November 26, 2000, p. 43.

Cohen, R., Erez, K., ben-Avraham, D., and Havlin, S. Resilience of the Internet to random breakdowns. *Physical Review Letters,* 85, 4626–4628 (2000).

———. Breakdown of the Internet under intentional attack. *Physical Review Letters,* 86, 3682–3685 (2001).

Coleman, J. S., Menzel, H., and Katz, E. The diffusion of an innovation among physicians. *Sociometry,* 20, 253–270 (1957).

Davis, J. A. Structural balance, mechanical solidarity, and interpersonal relations. *American Journal of Sociology,* 68(4), 444–462 (1963).

Davis, G. F. The significance of board interlocks for corporate governance. *Corporate Governance,* 4(3), 154–159 (1996).

Davis, G. F., and Greve, H. R. Corporate elite networks and governance changes in the 1980s. *American Journal of Sociology,* 103(1), 1–37 (1997).

Davis, G. F., Yoo, M., and Baker, W. E. The small world of corporate elite (working paper, University of Michigan Business School, 2002).

Degenne, A., and Forse, M. *Introducing Social Networks* (Sage, London, 1999).

De Vany, A., and Lee, C. Quality signals in information cascades and the dynamics of motion picture box office revenues: A computational model. *Journal of Economic Dynamics and Control,* 25, 593–614 (2001).

De Vany, A. S., and Walls, W. D. Bose-Einstein dynamics and adaptive contracting in the motion picture industry. *Economic Journal,* 106, 1493–1514 (1996).

Dodds, P. S., Watts, D. J., and Sabel, C F. The structure of optimal redistribution networks. Institute for Social and Economic Research and Policy Working Paper, Columbia University, (2002).

Durlauf, S. N. A framework for the study of individual behavior and social interactions. *Sociological Methodology,* 31, 47–87 (2001).

Ebel, H., Mielsch, L. I., and Bornholdt, S. Scale-free topology of e-mail networks. *Preprint* cond-mat/0201476. (2002). Available on-line at http://xxx.lanl.gov/abs/cond-mat/0201476.

Economides, N. The economics of networks. *International Journal of Industrial Organization,* 16(4), 673–699 (1996).

Emirbayer, M., and Mische, A. What is agency? *American Journal of Sociology,* 103(4), 962–1023 (1998).

Erdős, P., and Rényi, A. On random graphs. *Publicationes Mathematicae,* 6, 290–297 (1959).

———. On the evolution of random graphs. *Publications of the Mathematical Institute of the Hungarian Academy of Sciences,* 5, 17–61 (1960).

———. On the strength and connectedness of a random graph. *Acta Mathematica Scientia Hungary,* 12, 261–267 (1961).

Faloutsos, M., Faloutsos, P., and Faloutsos, C. On power-law relationships of the Internet topology. *Computer Communication Review,* 29, 251–262 (1999).

Fama, E. F. Agency problems and the theory of the firm. *Journal of Political Economy,* 88, 288–307 (1980).

Farmer, J. D. Market force, ecology, and evolution. *Industrial and Corporate Change,* forthcoming (2002).

Farmer, J. D., and Joshi, S. The price dynamics of common trading strategies. *Journal of Economic Behavior and Organization,* 49(2), 149–171 (2002).

Farmer, J. D., and Lo, A. Frontiers of finance: Evolution and efficient markets. *Proceedings of the National Academy of Sciences,* 96, 9991–9992 (1999).

Ferguson, N. M., Donnelly, C. A., and Anderson, R. M. The foot-and-mouth epidemic in Great Britain: Pattern of spread and impact of interventions. *Science,* 292, 1155–1160 (2001).

———. Transmission intensity and impact of control policies on the foot and mouth epidemic in Great Britain. *Nature,* 413, 542–548 (2001).

Ferrer i Cancho, R., Janssen, C., and Solé, R. V. Topology of technology graphs: Small world patterns in electronic circuits. *Physical Review E,* 64, 046119 (2001).

Flake, G. W. *The Computational Beauty of Nature: Computer Explorations of Fractals, Chaos, Complex Systems, and Adaptation* (MIT Press, Cambridge, MA, 1998).

Freeman, L. C. A set of measures of centrality based on betweenness. *Sociometry,* 40, 35–41 (1977).

Friedman, T. L. *The Lexus and the Olive Tree: Understanding Globalization* (Farrar, Straus and Giroux, New York, 1999).

Gell-Mann, M. *The Quark and the Jaguar: Adventures in the Simple and the Complex* (W. H. Freeman, New York, 1994).

Gibson, D., Kleinberg, J., and Raghavan, P. Inferring Web communities from link topology. In *Proceedings of the 9th ACM Conference on Hypertext and Hypermedia* (Association for Computing Machinery, New York 1998), pp. 225–234.

Gladwell, M. *The Tipping Point: How Little Things Can Make a Big Difference* (Little, Brown, New York, 2000).

Glaeser E. L., Sacerdote, B., and Schheinkman, J. A. Crime and social interactions. *Quarterly Journal of Economics*, 111, 507–548 (1996).

Glance, N. S., and Huberman, B. A. The outbreak of cooperation. *Journal of Mathematical Sociology*, 17(4), 281–302 (1993).

————. The dynamics of social dilemmas. *Scientific American*, 270(3), 76–81 (1994).

Glover, P. Grassroots economics. *In Context*, 41, 30 (1995).

Granovetter, M. Threshold models of collective behavior. *American Journal of Sociology*, 83(6), 1420–1443 (1978).

Granovetter, M. S. The strength of weak ties. *American Journal of Sociology*, 78, 1360–1380 (1973).

Grossman, J. W., and Ion, P. D. F. On a portion of the well-known collaboration graph. *Congressus Numerantium*, 108, 129–131 (1995).

Guare, J. *Six Degrees of Separation: A Play* (Vintage Books, New York, 1990).

Harary, F. Graph theoretic measures in the management sciences. *Management Science*, 387–403 (1959).

Harden, B. Dr. Matthew's passion. *New York Times Magazine*, February 18, 2001, pp. 24–62.

————. Coming to grips with the enduring appeal of body piercing. *New York Times*, February 12, 2002, p. A16.

Hardin, G. The tragedy of the commons. *Science*, 162, 1243–1248 (1968).

Hart, O. *Firms, Contracts and Financial Structure* (Oxford University Press, New York, 1995).

Hauer, J. F., and Dagel, J. E. *White Paper on Review of Recent Reliability Issues and System Events*. Prepared for U.S. Department of

Energy (1999). Available on-line at http://www.eren.doe.gov/der/transmission/pdfs/reliability/events.pdt.

Helper, S., MacDuffie, J. P., and Sabel, C. F. Pragmatic collaborations: Advancing knowledge while controlling opportunism. *Corporate Change, 9*, 3 (2000).

Hess, G. Disease in metapopulation models: Implications for conservation. *Ecology, 77*, 1617–1632 (1996).

Holland, J. H. *Hidden Order: How Adaptation Builds Complexity* (Perseus, Cambridge, MA, 1996).

Holland, P. W., and Leinhardt, S. An exponential family of probability distributions for directed graphs. *Journal of the American Statistical Association, 76*, 33–65 (1981).

Huberman, B. A., and Lukose, R. M. Social dilemmas and internet congestion. *Science, 277*, 535–537 (1997).

Ijiri, Y., and Simon, H. A. *Skew Distributions and the Sizes of Business Firms* (Elsevier/North-Holland, New York, 1977).

Jin, E. M., Girvan, M., and Newman, M. E. J. The structure of growing networks. *Physical Review E, 64*, 046132 (2001).

Kareiva, P. Population dynamics in spatially complex environments: Theory and data. *Philosophical Transactions of the Royal Society of London, Series B, 330*, 175–190 (1988).

Keeling, M. J. The effects of local spatial structure on epidemiological invasions. *Proceedings of the Royal Society of London, Series B, 266*, 859–867 (1999).

Keeling, M. J., Woolhouse, M. E. J., Shaw, D. J., Matthews, L., Chase-Topping, M., Haydon, D. T., Cornell, S. J., Kappey, J., Wilesmith, J., and Grenfell, B. T. Dynamics of the 2001 UK foot and mouth epidemic: Stochastic dispersal in a heterogeneous landscape. *Science, 294*, 813–817 (2001).

Kephart, J. O., Sorkin, G. B., Chess, D. M., and White, S. R. Fighting computer viruses. *Scientific American, 277*(5), 56–61 (1997).

Kephart, J. O., White, S. R., and Chess, D. M. Computer viruses and epidemiology. *IEEE Spectrum, 30*(5), 20–26 (1993).

Kermack, W. O., and McKendrick, A. G. A contribution to the mathematical theory of epidemics. *Proceedings of the Royal Society of London, Series A, 115*, 700–721 (1927).

———. Contributions to the mathematical theory of epidemics. II.

The problem of endemicity. *Proceedings of the Royal Society of London, Series A*, 138, 55–83 (1932).

———. Contributions to the mathematical theory of epidemics. III. Further studies of the problem of endemicity. *Proceedings of the Royal Society of London, Series A*, 141, 94–122 (1933).

Killworth, P. D., and Bernard, H. R. The reverse small world experiment. *Social Networks*, 1, 159–192 (1978).

Kim, B. J., Yoon, C. N., Han, S. K., and Jeong, H. Path finding strategies in scale-free networks. *Physical Review E*, 65, 027103 (2002).

Kim, H., and Bearman, P. The structure and dynamics of movement participation. *American Sociological Review*, 62(1), 70–94 (1997).

Kindleberger, C. P. *Manias, Panics, and Crashes: A History of Financial Crises*, 4th ed. (Wiley, New York, 2000).

Kleinberg, J. Authoritative sources in a hyperlinked environment *Journal of the ACM*, 46, 604–632 (1999).

———. The small-world phenomenon: An algorithmic perspective. In *Proceedings of the 32nd Annual ACM Symposium on Theory of Computing* (Association of Computing Machinery, New York, 2000), pp. 163–170.

———. Navigation in a small world. *Nature*, 406, 845 (2000).

———. Small-world phenomena and the dynamics of information. In Dieterich, T. G., Becker, S., and Ghahramani, Z. (eds.), *Advances in Neural Information Processing Systems, (NIPS)*, 14 (MIT Press, Cambridge, MA, 2002).

Kleinberg, J., and Lawrence, S. The structure of the web. *Science*, 294, 1849 (2001).

Kleinfeld, J. S. The small-world problem. *Society*, 39(2), 61–66 (2002).

Knight, F. H. *Risk, Uncertainty, and Profit* (London School of Economics and Political Science, London, 1933).

Kochen, M. (ed.). *The Small World* (Ablex, Norwood, NJ, 1989).

Kogut, B., and Walker G. The small world of Germany and the durability of national networks. *American Sociological Review*, 66(3), 317–335 (2001).

Korte, C., and Milgram, S. Acquaintance networks between racial groups—application of the small world method. *Journal of Personality and Social Psychology*, 15(2), 101 (1970).

Kosterev, D. N., Taylor, C. W., and Mittelstadt, W. A. Model validation for the August 10, 1996 WSCC system outage. *IEEE Transactions on Power Systems*, 14(3), 967–979 (1999).

Kretschmar, M., and Morris, M. Measures of concurrency in networks and the spread of infectious disease. *Mathematical Biosciences*, 133, 165–195 (1996).

Krugman, P. Fear itself. *New York Times Magazine*, September 30, 2001, p. 36.

Kuperman, M., and Abramson, G. Small world effect in an epidemiological model. *Physical Review Letters*, 86, 2909–2912 (2001).

Langewiescher, W. *American Ground: Unbuliding the World Trade Center* (North Point Press, New York, 2002).

Lawrence, S., and Giles, C. L. Accessibility of information on the web. *Nature*, 400, 107–109 (1999).

Liljeros, F., Edling, C. R., Amaral, L. A. N., Stanley, H. E., and Aberg, Y. The web of human sexual contacts. *Nature*, 411, 907–908 (2001).

Lohmann, S. The dynamics of informational cascades: The Monday demonstrations in Leipzig, East Germany, 1989–91. *World Politics*, 47, 42–101 (1994).

Longini, I. M., Jr. A mathematical model for predicting the geographic spread of new infectious agents. *Mathematical Biosciences*, 90, 367–383 (1988).

Lorrain, F., and White, H. C. Structural equivalence of individuals in social networks. *Journal of Mathematical Sociology*, 1, 49–80 (1971).

Lyall, S. Return to sender, please. *New York Times*, December 24, 2000, *Week in Review*, p. 2.

Lynch, N. A. *Distributed Algorithms* (Morgan Kauffman, San Francisco, 1997).

MacDuffie, J. P. The road to "root cause": Shop-floor problem-solving at three auto assembly plants. *Management Science*, 43, 4 (1997).

MacKenzie, D. Fear in the markets. *London Review of Books*, 22(8), 31–32 (2000).

Mackay, C. *Extraordinary Popular Delusions and the Madness of Crowds* (Harmony Books, New York, 1980).

Mannville, B. Complex adaptive knowledge management: A case study from McKinsey and Company. In Clippinger, J. H. (ed.), *The Biology of Business: Decoding the Natural Laws of the Enterprise* (Jossey-Bass, San Francisco, 1999), chapter 5.

March, J. G., and Simon, H. A. *Organizations* (Blackwell, Oxford, 1993).

Merton, R. K. The Matthew effect in science. *Science*, 159, 56–63 (1968).

Milgram, S. The small world problem. *Psychology Today* 2, 60–67 (1967).

———. *Obedience to Authority: An Experimental View* (Harper & Row, New York, 1974).

———. *The Individual in a Social World: Essays and Experiments*, 2nd ed. (McGraw-Hill, New York, 1992).

Milgrom, P., and Roberts, J. The economics of modern manufacturing: Technology, strategy, and organization. *American Economic Review*, 80(3), 511–528 (1990).

Mizruchi, M. S., and Potts, B. B. Centrality and power revisited: Actor success in group decision making. *Social Networks*, 20, 353–387 (1998).

Molloy, M., and Reed, B. A critical point for random graphs with a given degree sequence. *Random Structures and Algorithms*, 6, 161–179 (1995).

———. The size of the giant component of a random graph with a given degree sequence. *Combinatorics, Probability, and Computing*, 7, 295–305 (1998).

Monasson, R. Diffusion, localization and dispersion relations on 'small-world' lattices. *European Physical Journal B*, 12(4), 555–567 (1999).

Moore, C., and Newman, M. E. J. Epidemics and percolation in small-world networks. *Physical Review E*, 61, 5678–5682 (2000).

———. Exact solution of site and bond percolation on small-world networks. *Physical Review E*, 62, 7059–7064 (2000).

Moore, G. A. *Crossing the Chasm: Marketing and Selling High-Tech Products to Mainstream Customers* (Harper Business, New York, 1999).

Morris, S. N. Contagion. *Review of Economic Studies*, 67, 57–78 (2000).

Morse, M. Dollars or sense. *Utne Reader*, 99 (September–October 1999).

Murray, J. D. *Mathematical Biology*, 2nd ed. (Springer, Heidelberg, 1993).

Nadel, F. S. *Theory of Social Structure* (Free Press, Glencoe, IL, 1957).

Nagurney, A. *Network Economics: A Variational Inequality Approach* (Kluwer Academic, Boston, 1993).

Nelson, R. R., and Winter, S. G. *An Evolutionary Theory of Economic*

Change (Belknap Press of Harvard University Press, Cambridge, MA, 1982).

Newman, M. E. J. Models of the small world. *Journal of Statistical Physics*, 101, 819–841 (2000).

———. The structure of scientific collaboration networks. *Proceedings of the National Academy of Sciences*, 98, 404–409 (2001).

———. Scientific collaboration networks: I. Network construction and fundamental results. *Physical Review E*, 64 016131 (2001).

———. Scientific collaboration networks: II. Shortest paths, weighted networks, and centrality. *Physical Review E*, 64, 016132 (2001).

Newman, M. E. J., Barabási, A. L., and Watts, D. J. *The Structure and Dynamics of Networks* (Princeton University Press, Princeton, 2003).

Newman, M. E. J., and Barkema, G. T. *Monte Carlo Methods for Statistical Physics* (Clarendon Press, Oxford, 1999).

Newman, M. E. J., Moore, C., and Watts, D. J. Mean-field solution of the small-world network model. *Physical Review Letters*, 84, 3201–3204 (2000).

Newman, M. E. J., Strogatz, S. H., and Watts, D. J. Random graphs with arbitrary degree distributions and their applications. *Physical Review E*, 64, 026118 (2001).

Newman, M. E. J., and Watts, D. J. Scaling and percolation in the small-world network model. *Physical Review E*, 60, 7332–7342 (1999).

———. Renormalization group analysis of the small-world network model. *Physics Letters A*, 263, 341–346 (1999).

Newman, M. E. J., Watts, D. J., and Strogatz, S. H. Random graph models of social networks. *Proceedings of the National Academy of Sciences*, 99, 2566–2572 (2002).

Nishiguchi, T., and Beaudet, A. Fractal design: Self-organizing links in supply chain management. In Von Krogh, G., Nonaka, I., and Nishiguchi, T. (eds.) *Knowledge Creation: A New Source of Value* (Macmillan, London, 2000).

Noelle-Neumann, E. Turbulences in the climate of opinion: Methodological applications of the spiral of silence theory. *Public Opinion Quarterly*, 41(2), 143–158 (1977).

Nowak, M. A., and May, R. M. Evolutionary games and spatial chaos. *Nature*, 359, 826–829 (1992).

Olson, M. *The Logic of Collective Action: Public Goods and the Theory of Groups* (Harvard University Press, Cambridge, MA, 1965).

Ostrom, E., Burger, J., Field, C. B., Norgaard, R. B., and Policansky,

D. Revisiting the commons: Local lessons, global challenges. *Science, 284,* 278–282 (1999).

Palmer, R. Broken ergodicity. In Stein, D. L. (ed.), *Lectures in the Sciences of Complexity,* vol. I, Santa Fe Institute Studies in the Sciences of Complexity (Addison-Wesley, Reading, MA, 1989), pp. 275–300.

Pastor-Satorras, R., and Vespignani, A. Epidemic spreading in scale-free networks. *Physical Review Letters, 86,* 3200–3203 (2001).

———. Epidemics and immunization in scale-free networks. In Bornholdt, S., and Schuster, H. G. (eds.), *Handbook of Graphs and Networks: From the Genome to the Internet* (Wiley-VCH, Berlin, 2002).

Pattison, P. *Algebraic Models for Social Networks* (Cambridge University Press, Cambridge, 1993).

Perrow, C. *Normal Accidents: Living with High-Risk Technologies* (Basic Books, New York, 1984).

Piore, M. J., and Sabel, C. F. *The Second Industrial Divide: Possibilities for Prosperity* (Basic Books, New York, 1984).

Pool, I. de Sola, and M. Kochen. Contacts and influence. *Social Networks,* 1(r), 1–51 (1978).

Powell, W., and DiMaggio, P. (eds.). *The New Institutionalism in Organizational Analysis* (Chicago, University of Chicago Press, 1991).

Preston, R. *The Hot Zone* (Random House, New York, 1994).

Price, D. J. de Solla. Networks of scientific papers. *Science, 149,* 510–515 (1965).

———. A general theory of bibliometrics and other cumulative advantage processes. *Journal of the American Society of Information Science, 27,* 292–306 (1980).

Radner, R. The organization of decentralized information processing. *Econometrica,* 61(5), 1109–1146 (1993).

———. Bounded rationality, indeterminacy, and the theory of the firm. *Economic Journal,* 106, 1360–1373 (1996).

Rapoport, A. A contribution to the theory of random and biased nets. *Bulletin of Mathematical Biophysics,* 19, 257–271 (1957).

———. Mathematical models of social interaction. In Luce, R. D., Bush, R. R., and Galanter, E. (eds.), *Handbook of Mathematical Psychology,* vol. 2 (Wiley, New York, 1963), pp. 493–579.

———. *Certainties and Doubts: A Philosophy of Life* (Black Rose Press, Montreal, 2000).

Rashbaum, W. K. Police officers swiftly show inventiveness during crisis. *New York Times,* September 17, 2001, p. A7.

Redner, S. How popular is your paper? An empirical study of the citation distribution. *Europhysics Journal* B, 4, 131–134 (1998).

Ritter, J. P. Why Gnutella can't scale. No, really (working paper, available on-line http://www.darkridge.com/~jpr5/doc/gnutella.html, 2000).

Rogers, E. *The Diffusion of Innovations*, 4th ed. (Free Press, New York, 1995).

Romer, P. Increasing returns and long-run growth. *Journal of Political Economy*, 94(5), 1002–1034 (1986).

Sabel, C. F. Diversity, not specialization: The ties that bind the (new) industrial district. In Quadrio Curzio, A., and Fortis, M. (eds.), *Complexity and Industrial Clusters Dynamics Models in Theory and Practice* (Physica-Verlag, Heidelberg, 2002).

Sachtjen, M. L., Carreras, B. A., and Lynch, V. E. Disturbances in a power transmission system. *Physical Review E*, 61(5), 4877–4882 (2000).

Sah, R. K., and Stiglitz, J. E. The architecture of economic systems: Hierarchies and polyarchies. *American Economic Review*, 76(4), 716–727 (1986).

Sattenspiel, L., and Simon, C. P. The spread and persistence of infectious diseases in structured populations. *Mathematical Biosciences*, 90, 341–366 (1988).

Schelling, T. C. A study of binary choices with externalities. *Journal of Conflict Resolution*, 17(3), 381–428 (1973).

———. *Micromotives and Macrobehavior* (Norton, New York, 1978).

Scott, A. *Social Network Analysis*, 2nd ed. (Sage, London, 2000).

Shiller, R. J. *Irrational Exuberance* (Princeton University Press, Princeton, NJ, 2000).

Simon, H. A. On a class of skew distribution functions. *Biometrika*, 42, 425–440 (1955).

Simon, H. A., Egidi, M., and Marris, R. L. *Economics, Bounded Rationality and the Cognitive Revolution* (Edward Elgar, Brookfield, VT, 1992).

Smith, A. *The Wealth of Nations* (University of Chicago Press, Chicago, 1976).

Solomonoff, R., and Rapoport, A. Connectivity of random nets. *Bulletin of Mathematical Biophysics*, 13, 107–117 (1951).

Sornette, D. *Critical Phenomena in Natural Sciences* (Springer, Berlin, 2000).

Sporns, O., Tononi, G., and Edelman, G. M. Theoretical neuroanatomy: Relating anatomical and functional connectivity in graphs and cortical connection matrices. *Cerebral Cortex,* 10, 127–141 (2000).

Stanley, H. E. *Introduction to Phase Transitions and Critical Phenomena* (Oxford University Press, Oxford, 1971).

Stark, D. C. Recombinant property in East European capitalism. *American Journal of Sociology,* 101(4), 993–1027 (1996).

———. Heterarchy: Distributing authority and organizing diversity. In Clippinger, J. H. (ed.), *The Biology of Business: Decoding the Natural Laws of the Enterprise* (Jossey-Bass, San Francisco, 1999), chapter 7.

Stark, D. C., and Bruszt, L. *Postsocialist Pathways: Transforming Politics and Property in East Central Europe* (Cambridge University Press, Cambridge, 1998).

Stauffer, D., and Aharony, A. *Introduction to Percolation Theory* (Taylor and Francis, London, 1992).

Stein, D. L. Disordered systems: Mostly spin systems. In Stein, D. L. (ed.), *Lectures in the Sciences of Complexity,* vol. I, Santa Fe Institute Studies in the Sciences of Complexity (Addison-Wesley, Reading, MA, 1989), pp. 301–354.

Strogatz, S. H. *Nonlinear Dynamics and Chaos with Applications to Physics, Biology, Chemistry, and Engineering* (Addison-Wesley, Reading, MA, 1994).

———. Norbert Wiener's brain waves. In Levin, S. A. (ed.), *Frontiers in Mathematical Biology, Lecture Notes in Biomathematics, 100* (Springer, New York, 1994), pp. 122–138.

———. Exploring complex networks. *Nature,* 410, 268–275 (2001).

———. *Sync: The Emerging Science of Spontaneous Order* (Hyperion, Los Angeles, 2003).

Strogatz, S. H., and Stewart, I. Coupled oscillators and biological synchronization. *Scientific American,* 269(6), 102–109 (1993).

Travers, J., and Milgram, S. An experimental study of the small world problem. *Sociometry,* 32(4), 425–443 (1969).

Valente, T. W. *Network Models of the Diffusion of Innovations* (Hampton Press, Cresskill, NJ, 1995).

Van Zandt, T. Decentralized information processing in the theory of organizations. In Sertel, M. (ed.), *Contemporary Economic Issues,* vol. 4: *Economic Design and Behavior* (Macmillan, London, 1999), chapter 7.

Wagner, A., and Fell, D. The small world inside large metabolic networks. *Proceedings of the Royal Society of London, Series B*, 268, 1803–1810 (2001).

Waldrop, M. M. *Complexity: The Emerging Science at the Edge of Order and Chaos* (Touchstone, New York, 1992).

Walsh, T. Search in a small world. *Proceedings of the 16th International Joint Conference on Artificial Intelligence* (Morgan Kaufmann, San Francisco, 1999), pp. 1172–1177.

Ward, A., Liker, J. K., Cristiano, J. J., and Sobek, D. K. The second Toyota paradox: How delaying decisions can make better cars faster. *Sloan Management Review*, 36(3), 43–51 (1995).

Wasserman, S., and Faust, K. *Social Network Analysis: Methods and Applications* (Cambridge University Press, Cambridge, 1994).

Watts, D. J. Networks, dynamics and the small-world phenomenon. *American Journal of Sociology*, 105(2), 493–527 (1999).

―――. *Small Worlds: The Dynamics of Networks between Order and Randomness* (Princeton University Press, Princeton, NJ, 1999).

―――. A simple model of global cascades on random networks. *Proceedings of the National Academy of Sciences*, 99, 5766–5771 (2002).

Watts, D. J., Dodds, P. S., and Newman, M. E. J. Identity and search in social networks. *Science*, 296, 1302–1305 (2002).

Watts, D. J., and Strogatz, S. H. Collective dynamics of 'small-world' networks. *Nature*, 393, 440–442 (1998).

West, D. B. *Introduction to Graph Theory* (Prentice-Hall, Upper Saddle River, NJ, 1996).

White, H. C. What is the center of the small world? (paper presented at American Association for the Advancement of Science annual symposium, Washington, D.C., February 17–22, 2000).

White, H. C., Boorman, S. A., and Breiger, R. L. Social structure from multiple networks. I. Blockmodels of roles and positions. *American Journal of Sociology*, 81(4), 730–780 (1976).

Wildavsky, B. Small world, isn't it? *U. S. News and World Report*, April 1, 2002, p. 68.

Williamson, O. E. *Markets and Hierarchies* (Free Press, New York, 1975).

―――. Transaction cost economics and organization theory. In Smelser, N. J., and Swedberg, R. (eds.), *The Handbook of Economic Sociology* (Princeton University Press, Princeton, NJ, 1994), pp. 77–107.

Winfree, A. T. Biological rhythms and the behavior of populations of coupled oscillators. *Journal of Theoretical Biology*, 16, 15–42 (1967).

———. *The Geometry of Biological Time* (Springer, Berlin, 1990).

WSCC Operations Committee. *Western Systems Coordinating Council Disturbance Report, August 10, 1996* (October 18, 1996). Available on-line at http://www.wscc.com/outages.htm.

Zipf, G. K. *Human Behavior and the Principle of Least Effort* (Addison-Wesley, Cambridge, MA, 1949).

Index